Sauer

# The DC-Factory

Alexander Sauer

# The DC-Factory

Energy efficient. Robust. Forward-looking.

HANSER
Hanser Publishers, Munich

The Editor:
*Univ.-Prof. Dr.-Ing. Dipl.-Kfm. Alexander Sauer,* Head of the Institute for Energy Efficiency in Production (EEP) and the Fraunhofer Institute for Manufacturing Engineering and Automation (IPA) in Germany

MIX
Papier aus verantwortungs-
vollen Quellen
FSC® C083411

Distributed by:
Carl Hanser Verlag
Postfach 86 04 20, 81631 Munich, Germany
Fax: +49 (89) 98 48 09
www.hanserpublications.com
www.hanser-fachbuch.de


Editor: Dipl. Ing Volker Herzberg
Production Management: Cornelia Speckmaier
Coverconcept: Marc Müller-Bremer, www.rebranding.de, Munich
Coverdesign: Max Kostopoulos
Cover image: Johanna Nierich, using artwork by ©shutterstock.com/izabel.l
Typesetting: Eberl & Kœsel Studio GmbH, Altusried-Krugzell, Germany
Printed and bound by CPI books GmbH, Leck
Printed in Germany

ISBN: 978-3-446-47174-0
E-Book ISBN: 978-3-446-47179-5

# Contents

# Preface

The book “The DC-Factory. Energy efficient. Robust. Forward-looking.” gives a deep insight into the results of the research project DC-INDUSTRIE. It aims to motivate companies to rethink their own energy supply. A variety of arguments and incentives for this can be found in the following chapters.

Renewable energies usually produce direct current (DC). This is initially converted into alternating current (AC) and fed into the grid. Power electronic devices in the factory then convert this back into direct current, for example to drive electric motors. Storage solutions usually work with direct current, too. Transformations from direct current to alternating current and vice versa, however, damage the quality of supply due to the rapid switching of power semiconductors. In industrial companies this can lead to EMC problems and thus to production damage.

On the other hand, the characteristics of production in DC grids are manifold: high quality, efficient, resource-saving, robust, and inexpensive.

A DC factory, for example, does not need conversion stages in frequency converters or in PV inverters. It saves on components and the remaining devices can be smaller. There is also a lot to be saved with respect to cables. Unlike the industrial 3-phase AC grid, a DC grid is 2-phase. This means that one, sometimes even two conductors in the cabling are no longer necessary. A further efficiency effect is that the braking energy of electric motors can be fed back into the grid more easily.

As a smart grid, today's DC grid can split the power flow between different storage devices and consumers. This makes the supply of direct current particularly robust. With low process dynamics, a battery can reduce the connected load; with higher dynamics, a capacitor can compensate for short-term power boosts. If a lot of power is needed, a flywheel mass storage can be added. The interaction of these technologies was successfully implemented and tested in the DC-INDUSTRIE research project.

The factory's power supply in the DC grid will be even more cost-effective if time-flexible energy tariffs can be used. This is because the system can then change the operating points of the storages and of the controllable consumers by means of a higher-level grid control. In this way, for example, the degree of utilization of the self-generated energy is influenced.

This book on the DC factory addresses all the important questions that arise when the voltage type is changed from AC to DC. Experts explain in detail – and yet easily understandable for the interested layperson – how a DC factory works, what its advantages are and how it can be implemented.

I would like to personally thank all the companies involved and especially the many dedicated experts for their work on the first implementation of an open DC grid in industrial plants and for their support in the preparation of this book. Special thanks go to the Federal Ministry of Economics and Energy (BMWi), which supported the work, as well as to the *Heinz and Heide Dürr Foundation* and the *Karl Schlecht Foundation* for financing this book.

Stuttgart, July 2020

*Alexander Sauer*

Head of the Institute for Energy Efficiency in Production (EEP) and the Fraunhofer Institute for Production Engineering and Automation (IPA)

Climate change is currently one of the greatest challenges facing humanity. In order to be able to counteract climate change, it is necessary to drastically reduce global $CO_2$ emissions. This can only be achieved by decarbonizing the current energy production landscape. The expansion and integration of renewable energies into the existing power grid will make a major contribution to this. However, this is accompanied by the great challenge, especially for industrial companies, to be able to react optimally to fluctuating energy supplies and the associated lower quality of energy supply without letting this negatively affect the underlying production process.

In addition to the decarbonization of energy production, increasing efficiency within the numerous energy conversion chains plays a decisive role on the way to a more sustainable and environmentally friendly industrial landscape. Especially considering that in Germany, for example, approximately 45% of electrical energy consumption is accounted for by industry alone.

The DC-INDUSTRIE research consortium, funded by the German Federal Ministry of Economics and Technology (BMWi), has devoted three years of intensive work to the above-mentioned challenges with the development of the world's first open industrial DC grid for factory automation. The primary goal of the project was to ensure a demand-oriented distribution of energy within production plants with a maximum of energy reuse and a minimization of conversion losses. This work has resulted in the present book, which aims to take the reader on our joint "development journey".

The first part of the book discusses the potential of an industrial DC power supply and the resulting opportunities for drive technology.

In the second part, starting with a requirements analysis, the DC-INDUSTRIE system concept is presented, which – if the specifications agreed upon are met – allows a manufacturer-independent operation of different devices and the simple integration of renewable energy sources and suitable storage systems (batteries, capacitor banks, flywheel energy storages) on a common DC grid. In addition, a cross-manufacturer, decentralized grid management concept for highly dynamic industrial DC microgrids is presented.

In the third and last part of the book, various aspects of safety and protection technology are discussed, particularly with regard to the requirements on and design of DC circuit breakers developed specifically for this purpose. In addition, the reader is guided through a planning and design process that has been specially developed for the novel open DC grid presented here. Finally, the successful implementation and validation of the DC-INDUSTRIE concept in four different demonstration plants is presented.

This foreword would be very incomplete if I would not say a few words about the great commitment of the DC-INDUSTRIE consortium, which made this work possible in the first place.

The prerequisite for the extraordinarily strong innovative power of the project was, in addition to the concentrated technical and interdisciplinary competence of the consortium, a team spirit characterized by trust, respect, transparency, openness, commitment and an open feedback culture.

I would like to emphasize that it would have been contradictory and more than negligent to believe that it would have been possible to jointly design an open system concept without being able to presuppose an openness in thinking and acting on the part of each individual.

On the technical side, it was more important than ever to bundle the competencies of all necessary disciplines in daily work and to always keep the overall system in view in order to avoid conducting isolated or "sterile" research on individual technologies.

Finally, I would like to take this opportunity to thank the German Federal Ministry of Economics and Energy (BMWi) on behalf of all DC-INDUSTRIE team members and partner companies for the support of the project and the trust placed in us.

A very big thank you also goes to all DC-INDUSTRIE steering committee members, represented by their spokesman, Prof. Dr. Alexander Sauer, for their valuable advice and continuous support throughout the project.

My thanks would be very incomplete if I did not include in it the scientific leader (and one of the initiators) of the consortium, Prof. Dr. Holger Borcherding, for his extraordinary foresight and his continuous support, which have greatly shaped the project.

Erlangen, May 2020

*André Leonide*

Consortium Project Manager DC-INDUSTRIE

# 1 Direct Current Returns

Alexander Sauer
Karl-Peter Simon
Sebastian Weckmann

At the end of the 19th century, the age of electricity began. At that time, outstanding engineers, inventors and entrepreneurs were striving to develop a power supply system that would ensure both power generation and the transmission and distribution of power. However, what also flared up with the beginning of the electricity age was a battle for supremacy between direct current and alternating current technology (abbrev. DC and AC). It went down in history as the "battle of the currents".

With the discovery of the dynamoelectric principle in 1866 by Werner von Siemens, larger quantities of electrical energy became available for the first time [Siem1891]. This energy could be transported over long distances without great losses owing to the invention of the transformer in 1881 by Lucien Gaulard and John Dixon Gibbs [Walt2005]. Transformed to a higher voltage, electrical power could from now on be transmitted with lower current and thus lower line losses. In combination with the invention of two-phase alternating current in 1887 by Nikola Tesla and three-phase alternating current in 1888 by Dolivo Dobrowolski, these technologies were so successful that they still form the basis of our electricity system today [Grave1921].

In the early days, smaller island grids were used to supply electrical lighting and smaller DC motors in the 110 Volt range. Edison, the inventor of direct current technology, knew the disadvantage of the smaller transmission distance of his 110 Volt direct current networks compared to highly transformed alternating current and wanted to compensate for it by a large number of local smaller power plants [Cowd2006].

Edison's patent for incandescent lamps from 1880 gave a decisive boost to the development of electricity. He realized that he could only sell his lamps if an electricity network was available to the public. He bought an old building on Pearl Street in New York City and turned it into a power station. In September 1882, 800 lamps were used to light various buildings among which was the headquarters of the New York Times. After this remarkable success, Edison founded the General Electric Company. Fascinated by the burgeoning electrical power business, business magnate George Westinghouse recognized the weaknesses in Edison's DC system in terms of transmission efficiency. He purchased patent rights for the transformer and to Tesla's AC motor. By 1889, Westinghouse had built 870 AC grids thereby limiting Edison's profits. Edison was also concerned about the social acceptance of electricity use. He feared that his business model might be jeopardized by possible accidents at high voltages in the alternating current networks. In particular, the safety of electricity – compared to the fire hazard of gas – was one of his central arguments [Isra2000]. In order to discredit alternating current, Edison proposed it for the execution of the death penalty. To demonstrate its effectiveness, Edison's associate Harold Brown traveled from city to city, publicly killing dogs, cows, horses, and even a three-ton elephant named "Topsy" [Cowd2006].

Ultimately, there were two historical milestones that led to the breakthrough of alternating current technology. At the Chicago World Fair in 1893, the first all-electric world exhibition, 27 million visitors experienced the illumination with 100,000 incandescent lamps. The most spectacular lighting the world had ever seen. In 1890, the International Niagara Commission, led by Lord Kelvin, funded a competition to harness the energy of the Niagara Falls. Originally Kelvin preferred a direct current grid but changed his mind after attending the Chicago Fair. The decision was taken to use alternating current technology. At the time, alternating current technology was the preferred choice because it was about half the price of direct current technology. This was mainly due to the direct transmission of the three-phase alternating current produced by generators in combination with passive transformers for scaling the voltage level [Cowd2006].

Since the Chicago World Fair, the world we live in and the energy system associated with it has changed fundamentally. In 1947, the scientists Bardeen, Brattain and Shockley developed a new electronic component that could control the flow of current by means of an electrical signal – the transistor. Around the same time special semiconductor components were developed: above all the diode, but also the thyristor and the triac, which are suitable for switching currents of several thousand amperes [Klos1987]. The transistor brought new possibilities for the production of integrated circuits. In particular the military and the associated space travel made ever higher demands on electronic components in terms of reliability, performance, volume and weight. Demands that could only be met by integrated circuits. In the course of the development of microprocessors, semiconductor technology found its way into public life everywhere. Today there is hardly a technical device without semiconductor electronics [Fair2012a]. All these electronic circuits require DC voltage. Direct current technology has the potential to replace the alternating current technology sooner or later, because it is more efficient and cheaper.

The number of direct current consumers in our daily life is constantly increasing. LEDs (light emitting diodes), which are a major trend in the lighting industry, are only one example. LED lights have a high energy efficiency and a life span that is ten times longer than that of energy-saving lamps. With an operating time of eight hours per day, LEDs last on average 20 years. Only after that is it necessary to replace them with a new one. LEDs work with direct current. If – as for example in data centers – not every lamp is equipped with an inverter (i.e., an AC/DC converter), instead a local DC network with a central inverter is set up, the overall efficiency (i.e., the minimization of power loss) and resource efficiency can be improved, because today's AC/DC converters have a life expectancy that is significantly lower than that of LEDs. In addition, advanced lighting controls are being developed that are fully integrated into the DC system at no extra cost [Gago2018].

Mobile and Internet-enabled devices like smartphones and tablets are changing the way we communicate with each other today. A basic requirement is that users can connect their devices to a DC power source via a charging cable in order to charge the integrated electrical energy storage device with direct current. State of the art local chargers are used, each of which converts from a power outlet (230 Volt AC) to the required DC voltage (5 Volt DC), thereby generating electricity heat losses. The same applies to integrated power supplies in PCs, monitors and televisions.

Consumers such as heating pumps, air conditioners and fans are parts of building automation. Depending on their output, they also need a DC low voltage or a single-phase frequency converter with a DC intermediate circuit (approx. 325 V).

Data centers, which are necessary for the Internet and our telecommunications networks, nowadays consume more than 1.3% of the world's electricity – and this figure is rising. Instead of installing AC/DC converters in every server, some companies are using large, centralized inverters distributing 380 volt DC in their server farms. Energy savings are achieved primarily by replacing the AC/DC converters integrated in individual servers with more efficient central rectifiers. Switching to these central rectifiers and a more efficient connection of battery backup systems reduces power consumption by 15% compared to conventional AC configurations [Fair2012b].

The spread of electromobility makes direct current technology even more important. Many vehicles are charged with alternating current because it is available at home, at shopping locations or at work. The conversion to direct current is then done in the vehicle itself. However, the installation space for the respective converters is limited due to reasons of cost, space and weight. This means that – depending on the vehicle – it can take between four and over twelve hours to fully charge the battery. DC rapid charging systems bypass the limitations of in-vehicle converters. They can be used to significantly increase the charging speed. DC rapid charging is essential for journeys with high mileage and for large fleets. A fast turnaround allows the driver to charge during the day or during a short break [Schr2012].

Massive changes in the electricity system can also be witnessed on the generation side: the expansion of renewable energies is one of the major development issues worldwide. In Germany it is driven by various policies collectively referred to as the energy turnaround, the "Energiewende". As early as 2007, the EU decided that the share of renewable energies in the gross final energy consumption (electricity, heat, fuels) of the EU member states should rise to 20% by 2020. In addition, in 2018 the EU presented its long-term strategic vision for a prosperous, modern, competitive and climate-neutral economy by 2050. For Germany, a share of renewable energies of 18% of the primary energy demand is the objective to be met by 2020. By 2030 30%, by 2040 45% and by 2050 60% should be reached [Ren2019]. Overall, renewable energies now account for more than 40% of installed capacity [UBA2020]. The DC voltage that solar power plants produce anyway is converted into AC/DC current via DC/AC converters. The three-phase current produced by wind power plants (rotation frequency proportional to the rotational speed of the blades) is also first buffered in a DC intermediate circuit and then extracted again as three-phase current (rotation frequency proportional to the three-phase grid). This intermediate conversion is necessary in order to be able to feed into the existing AC grid in a synchronized way at 50/60 Hertz.

Energy storage is one of the main factors with regard to achieving energy flexible systems because of the fluctuating energy generation and existing grid bottlenecks. The ability to store locally generated energy and use it later is of great importance. Storage is a solution to the problems of unstable or expensive power grids and it is a crucial incentive for increasing the use and integration of renewable energies. There is a wide range of technologies for storing electricity: accumulators, gas, compressed air and chemicals, and – last, but not least – pumped storage power plants. The growing need for flexibility and autonomy, in which energy consumers also become energy producers, is driving more and more innovations in energy storage, resulting in a wide range of new solutions. The most important storage technologies for commercial and industrial plants are accumulators for storing electrical energy or thermal energy storage systems for storing cold water, ice or heat. These can be operated as stand-alone systems or in collaboration with solar systems. Because of the underlying physical operating principles, the electrical storage systems are almost exclusively operated with direct current [Chio2018].

Power transmission technologies have also evolved. For example, high-voltage direct current transmission technology offers several advantages over alternating current transmission systems because it enables a more efficient transmission of mass power over long distances. If the transmission line is longer than about 480 km, direct current is the better option because alternating current lines have higher line losses [Rai2016].

Direct current is also used in industrial production. In production facilities such as injection molding, machine tool and packaging machines or in industrial robots, motors are driven on demand by servo or frequency converters as electronically variable-rotation speed drives. Within these converters, the AC voltage of the supply network is typically first converted with rectifiers into an intermediate circuit DC voltage in order to then – with the help of a self-commutated inverter – be able to provide an AC voltage for the motor that can be changed in amplitude and frequency.

The simplest and most common form of rectification is by uncontrolled diode rectifiers. These generate high pulse-shaped recharging currents at the voltage maxima.

The increasing use of converters, which – due to the input rectifiers – load the grid with very high non-sinusoidal currents, leads to a decrease of the overall grid quality. Ever more frequently, this leads to a deterioration of the voltage quality to a level that requires considerable filtering measures at the feed-in point of a factory in order not to interfere inadmissibly with the external AC grid. These retroactive grid perturbations are caused because a large number of electrical consumers (such as power supply units or converters) are in operation at a high switching frequency thereby influencing the stability of the power grid. Such retroactive effects can be voltage dips (caused by transient switching operations) or voltage distortions (caused by non-linear consumers). The systems for voltage filtering and reactive power compensation usually require passive components that are not only space-intensive and expensive, but also require the use of valuable resources. See also the explanations in chapter 4 "Opportunities for Drive Technology".

At the same time, each conversion from alternating to direct current and vice versa causes considerable conversion losses; as already mentioned, this also applies to LED lights, chargers for smart devices as well as for data centers. Thus, the increasing use of variable-speed drives favors the consideration of an alternative network structure in which DC voltage is available everywhere within a factory as a distribution grid.

In 2016, the final energy consumption in Germany for industry, commerce, trade and services was 1128 TWh. This corresponds to around 45% of total German final energy consumption. Since 2008, the final energy productivity (energy efficiency) of the German economy has increased by more than 10%, but at the same time the absolute final energy consumption remained almost constant with a change of −0.1%. Therefore, further considerable efforts are essential to achieve national and international climate protection goals. Without a drastic improvement in energy efficiency, no economic energy turnaround is possible. However, this means – among other things – a significant increase in the use of variable-speed drive systems. EU legislation (19 Commission Regulation (EU) 2019/1781 of October 1, 2019, laying down ecodesign requirements for electric motors and variable speed drives) does not yet promote this system approach, as only individual components continue to be regulated by law. Thus, the focus is not on optimizing the entire application in terms of energy efficiency and economics. A drastic rethinking of the system approach in conjunction with a change in the network structure is necessary if the efficiency goals are to be achieved.

Direct current technology is more present today than ever before. For Europe alone, a market growth of 1000% is predicted for the period from 2017 to 2025 [Fros2020]. Against the backdrop of increasing cost pressure and the social demand for ever greater savings in terms of resources and $CO_2$ emissions, industry in particular is forced to significantly increase its energy productivity [Gerb2018].

The general conditions have changed: Direct current is generated decentrally directly from solar and wind energy. Today, direct current can also be efficiently transmitted over long distances. In an industrial DC grid, the negative repercussions on the grid quality are eliminated, conversion losses are reduced, and the recovery of regenerative kinetic energy (recuperation) becomes an integral part.

Direct current technology is the essential key to a new energy age. So far, direct current has only been used on an industrial scale to supply production lines in proprietary areas. However, these self-contained subnetworks cannot develop the potential of an entire factory. The following chapters show how an open direct current network can be designed, which technologies would be necessary, how they can be used and where the specific challenges lie.

## Literature


[Siem1891] Siemens, Werner: Die dynamo-elektrische Maschine. Springer-Verlag, Berlin, Heidelberg, 1891

[Walt2005] Walter, H.-J.: Die Erfinder des Transformators. In: Elektrische Maschinen, Huthig Verlag, p. 28., 2005

[Grab1921] Grabscheid, Johann: Elektromotoren. Springer-Verlag, Berlin, Heidelberg, 1921

[Cowd2006] Cowdrey, John: The war of the currents. Home Power Magazine 111. Oregon, 2006, http://h2oradio.org/PDF/WaroftheCurrents_Cowdrey.pdf (access on 23 Apr 2020)

[Isra2000] Israel, Paul: Edison: A Life of Invention. John Wiley & Sons-Verlag, New York, 2000

[Klos1987] Kloss, Albert: Von der Electricität zur Elektrizität. Birkhäuser-Verlag, Basel, 1987

[Fair2012a] Fairley, Peter: DC Versus AC: The Second War of Currents Has Already Begun [In My View]. *IEEE Power and Energy Magazine*. Institute of Electrical and Electronics Engineers, New York, 2012

[Gago2018] Gago-Calderón, Alfons; Orejón-Sánchez Rami D.; Hermoso-Orzácz, Manolo J.: DC Network Indoor and Outdoor LED Lighting. IntechOpen-Verlag, London, 2018

[Fair2012b] Fairley, Peter: Edison's Revenge: The Rise of DC Power. Technology Review Magazine. Massachusetts Institute of Technology, Cambridge, 2012

[Schr2012] Schroeder, Andreas; Traber, Thure: The economics of fast charging infrastructure for electric vehicles. Elsevier-Verlag, Berlin, 2012

[Ren2019] N.N.: Renewables 2019 Global Status Report. REN21, Paris, 2019, https://www.ren21.net/wp-content/uploads/2019/05/gsr_2019_full_report_en.pdf (access on 23 Apr 2020).

[UBA2020] N.N.: Erneuerbare Energien in Zahlen. *https://www.umweltbundesamt.de/themen/klima-energie/erneuerbare-energien/erneuerbare-energien-in-zahlen#uberblick* (access on 01 Apr 2020)

[Chio2018] Chiodo, Elio; Fantauzzi, Maurizio; Lauria, Davide; Mottola, Fabio: A Probabilistic Approach for the Optimal Sizing of Storage Devices to Increase the Penetration of Plug-in Electric Vehicles in Direct Current Networks. Energies. Multidisciplinary Digital Publishing Institute, Basel, 2018

[Rai2016] Rai, Anil Kumar; Sharma, Chandra Shekhar: DC Vs AC War Of Currents For Future Power Systems A HVDC Technology Overview. International Journal of Scientific & Technology Research, Delhi, 2016, *https://www.ijstr.org/final-print/may2016/Dc-Vs-Ac-War-Of-Currents-For-Future-Power-Systems-A-Hvdc-Technology-Overview.pdf* (access on 23 Apr 2020)

[Fros2020] N.N.: European DC Power Distribution Market, Forecast to 2025: Collaboration amongst OEMs, Distributed Energy Providers, and Governments is Critical for Implementing the DC Infrastructure. Frost & Sullivan, New York, 2018, https://store.frost.com/european-dc-power-distribution-market-forecast-to-2025.html (access on 01 Apr 2020)

[Gerb2018] Gerbert, Philipp; Herhold, Patrick; Burchardt, Jens; Schönberg, Stefan; Rechenmacher, Florian; Kirchner, Almut; Kemmler, Andreas; Wünsch, Marco: Klimapfade für Deutschland. Boston Consulting Group, München, 2018, https://bdi.eu/publikation/news/klimapfade-fuer-deutschland/ (access on 27 Apr 2020)

# 2 Potentials of an Industrial DC Supply

Timm Kuhlmann
Patrick Spanier
Martin Ehlich

## ■ 2.1 Change from an AC to a DC Grid

AC grids are state-of-the-art in all factories worldwide. There are major regional differences in terms of voltage level, frequency and earthing systems. For this reason, AC devices are generally developed and qualified today for variable input voltages and different connection conditions.

### Consumer in an AC Grid

In the industrial sector, three-phase electric drives with a DC intermediate circuit (400 V...800 V) are the "driving" force in all machines and systems. They convert over 70% of the electrical energy used in a factory into mechanical motion.

A present-day factory with typical interconnected grid participants is shown in Figure 2.1.

The factory-internal AC distribution is usually fed via a separate central transformer at the grid connection point. There may be several such feed-in points if either the power is insufficient, or it requires redundancy for safety reasons. The AC distribution works in a 3-phase mode. Individual areas, machines, or larger individual consumers can be switched on and off separately via contactors. In the event of a fault, a circuit breaker or fuse separates these energy-using zones from the AC grid. Figure 2.1 does not show these switching and protection elements due to simplification. The focus is on the typical consumers within these zones:

- three-phase motors directly connected to the AC grid, e.g., in pumps, fans, and air conditioners or in use to generate compressed air and hydraulics
- individual processing machines and robots with many individually controllable axes
- position-, rotation speed- or torque-controlled individual drives as found in all production lines as well as in conveying and lifting applications

- welding cases for resistance welding in the automotive industry; these also use converters with a DC intermediate circuit to control the primary side of the medium-frequency welding transformer
- passive consumers, such as heaters for heating furnaces, soldering systems in electronics production or heating applications for plastics or adhesive processing
- all types of automation equipment, such as machine and system controls, communication systems, proximity switches, light barriers, etc. require a DC voltage of 24 V to supply the integrated electronics; this auxiliary power supply is provided by a variety of separate or device-integrated AC power supplies
- sockets for portable devices, electrical tools for local maintenance or assembly work or measuring devices; 230 V sockets are often provided in the switch cabinets as well

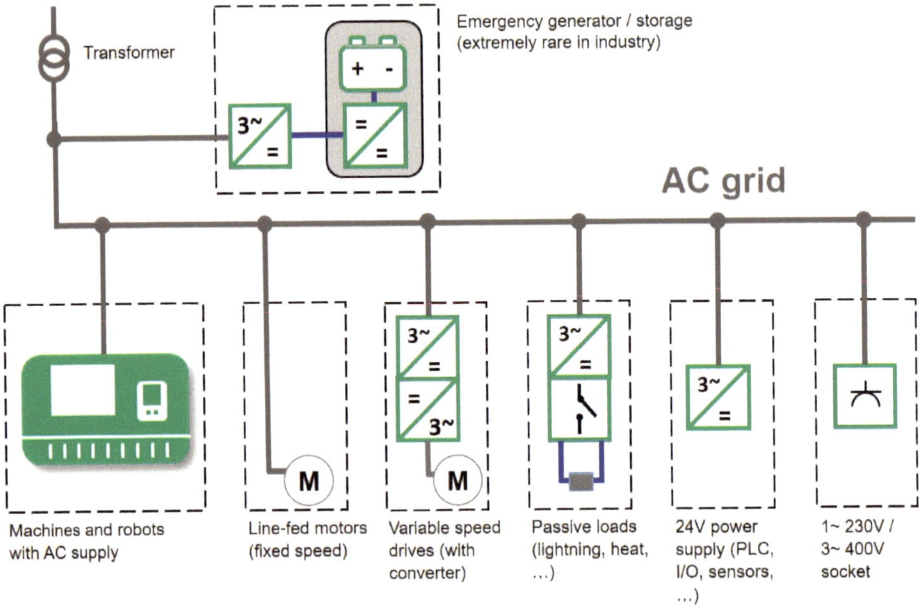


**Figure 2.1** Topology in an industrial AC factory grid

### Regeneration in an AC Grid

In principle, the current direction in AC distribution is bidirectional, which is actually used in some applications. However, only those consumers can feed back energy, where generative energy is produced in the process. This energy has to be converted accordingly so that it can be output synchronously to the frequency of the AC grid. This is only possible with actively controlled input inverters, which are much more complex and expensive than the most commonly used unidirectional diode rectifiers. Therefore, energy is only fed back into an AC grid if a significant reduction in energy costs covers the investment.

### Power Failures in an AC Grid

In the event of a power failure, the entire production process is stopped immediately. This is often the case even with very short interruptions of less than one second. From a technical point of view, there is the possibility to bridge a certain period of power failure with the help of battery storage. However, the costs for acquisition, installation, maintenance and installation space are so high that such systems are usually only used in security-relevant areas such as nuclear power plants, data centers and hospitals.

### Reasons for Switching to a DC Grid

An analysis of industrial consumers shows that in all application areas there is an increasing number of interconnected devices that convert the input AC voltage internally into a DC voltage anyway. The battery storage units for bridging AC power failures can only be supplied with DC voltage. The idea of supplying these consumers directly with DC voltage is obvious.

At the same time, there are more and more energy generators that generate a direct current that can be fed directly into a DC voltage grid via DC/DC converters (Figure 2.2). These include the most widespread – since they can be used decentrally practically everywhere – regenerative technologies: wind energy and photovoltaic (PV) systems.

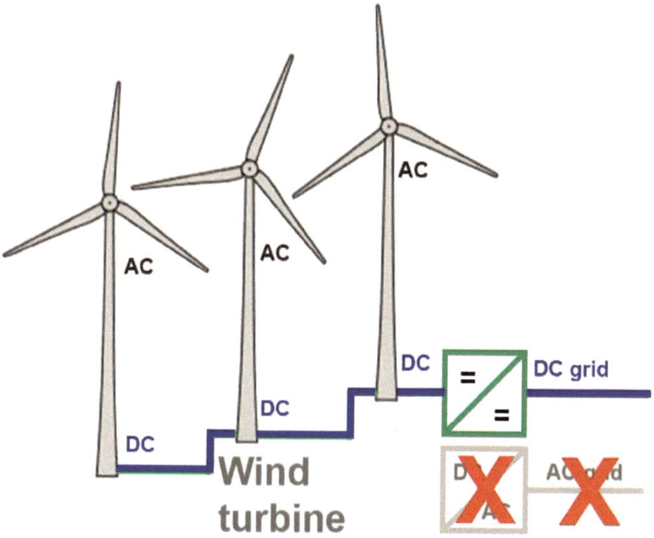


**Figure 2.2** Wind turbine, directly at the production site for green energy

In wind turbines, the three-phase voltage system generated by rotation is firstly converted into a DC voltage with the aid of electronic AC/DC converters because the generators initially generate a grid-asynchronous AC voltage. The inverter (DC/AC converter) required today can be replaced by a relatively simple DC/DC converter for connecting to a DC grid. This allows locally generated wind energy to be used directly in a DC factory grid.

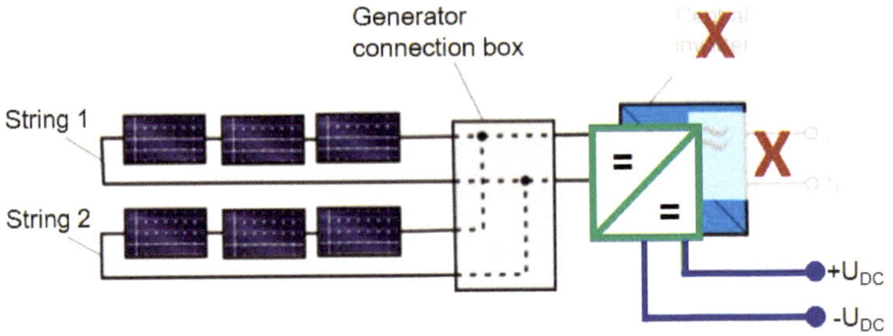


**Figure 2.3** Photovoltaic system for green energy directly from the factory roof

In principle, PV systems generate a DC voltage. The level of the DC voltage depends on the interconnection of the solar modules (number of serially connected modules in a string) (Figure 2.3). For use in a classical AC grid, an inverter is also required here, which feeds the current into the grid synchronously and in the correct phase. The connection of a PV system to a DC grid is easier to implement.

Owing to the high importance and the widespread distribution of variable rotation speed drives with a DC intermediate circuit, there have been efforts for a long time to establish an energy exchange at least in the local drive grid. Many manufacturer-specific solutions already make this possible for limited, proprietary interconnected systems. However, they do not offer an open, manufacturer-independent standard for entire plants and factories. These so-called DC link grids already use all the advantages of a DC grid on a small scale. The manufacturers and end users of such drive axles know exactly that in a dynamic multi-axle system a lot of recuperated energy is generated that could be reused. Saving one rectifier and one braking resistor per axis reduces equipment costs and installation space.

Other non-industrial areas use DC distribution grids where compatibility or connection to an AC grid is not necessary. These include, for example, on-board power supplies in electric vehicles and in ships. The charging infrastructure for electric vehicles is partially operated on a DC distribution grid. This is the case, when the DC distribution grid extends beyond individual charging points, thus eliminating the need for multiple conversions from AC to DC. In larger data centers it is already state-of-the-art to supply all units via a DC distribution grid.

### The Idea of an Open DC Grid

The comprehensive idea is to supply all the systems mentioned so far - as far as they occur in the industry - continuously via a common, open, i.e., manufacturer-independent DC grid. Such an industrial DC grid should be scalable from individual machines to entire factory floors (buildings). All advantages of DC technology can be used without the need for changes in the external supply lines or at the medium voltage level.

Due to the system-related direct energy exchange between motorized and generative processes, a self-regulating energy balance results in the DC grid. This means that only the energy required for the implementation of production processes, minus any integrated renewable energy sources, can be obtained from the AC grid connection. This energy is mainly converted in machining, processing and conveying processes. Ideally, superfluous energy from motion sequences should no longer be burned at braking resistors.

An open DC grid with typical and possible grid participants is sketched out in Figure 2.4.

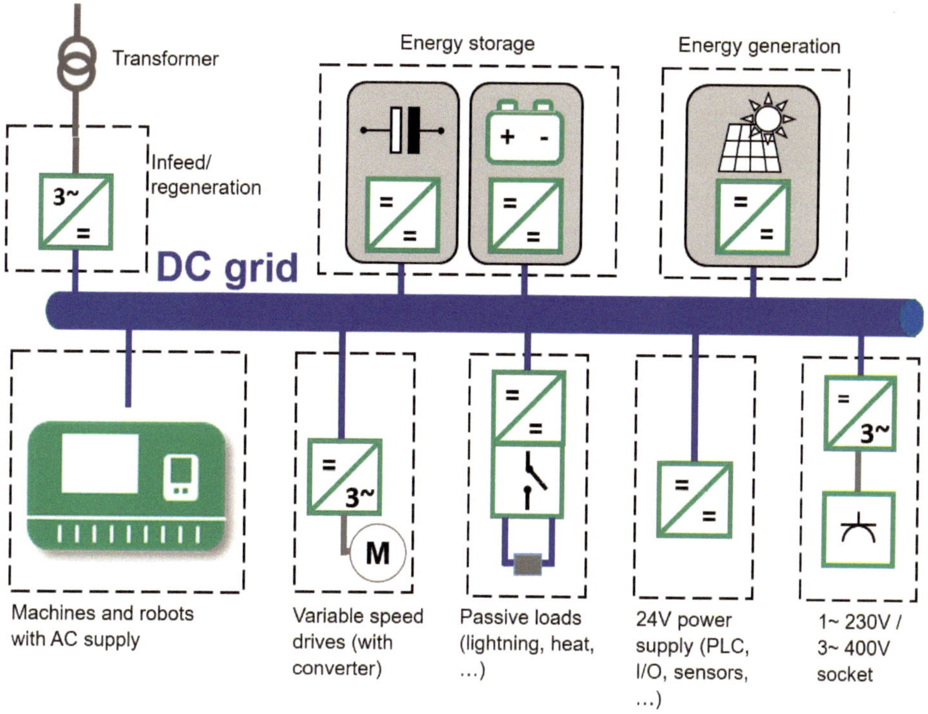


**Figure 2.4** Topology in an industrial DC factory grid

In the direct current factory, the same processes and motion sequences must be fulfilled as before. In contrast to the AC factory grid, there are one or more central supply units as AC/DC converters, which provide the energetic connection to the "feeding" AC grid. Depending on the design of the devices, the energy flow would be unidirectional (pure feed, e.g., via diode rectifiers) or bidirectional (feed-in and feedback with the help of rectifiers and inverters).

Other features of the DC factory grid sketched out here are:

- All drives in the machines and robots are directly connected to the DC grid. This enables energy to be exchanged not only within a group of axles, but also between different parts of the plant.
- Grid-synchronous drives are replaced by variable rotation speed drives. This simplifies the topology within the inverters. The input rectifier is omitted, the EMC filter is simplified by the omission of the line reactor, and the required DC link capacity is reduced (see also Chapter 4.2 and Figure 4.5). This reduces costs, energy and space requirements. Furthermore, it is easier to integrate the electronics into the motor, thus favoring the use of decentralized drives. These in turn offer the following advantages over control cabinet devices:
  - no control cabinet
  - no expensive, lossy motor cables
  - reduced reloading losses
  - simple line wiring
  - modular machines
- Passive loads are controlled via DC/DC converters instead of the AC/DC converters previously required.
- The DC power supply units for providing the auxiliary power supply are simpler, more efficient and cheaper than comparable AC power supply units because the internal rectification is omitted and the effort for filtering is reduced, as can be seen with the frequency converters (see Chapter 4.2 and Figure 4.5).
- For the connection of mobile 3-phase and 1-phase AC consumers, which are currently not yet available with DC connection, local connections (sockets as "AC island grids") can be provided via DC/AC converters.
- Static energy storage devices in the form of batteries or capacitors can easily be coupled to the DC grid. Capacitor banks can be coupled directly to the DC grid, for example to intercept local peak loads and to stabilize the DC grid. Battery banks are connected via DC/DC converters for voltage adjustment and power control. Storage capacity and storage technology can be designed to bridge short-term AC grid disturbances up to medium AC grid downtimes.

- Regenerative energy from PV systems or wind power plants can be fed into the DC grid using DC/DC converters. If this energy is not currently being used within the factory, it can be fed into the AC grid and sold via the feed-in/feedback system.

The advantages of the DC grid compared to a comparable system using AC technology are:

- lower energy consumption at same production output levels
- ideally suited for the integration of green energy sources
- round-the-clock availability is possible with the help of modern energy storage systems
- reduced grid connection capacity with correspondingly lower electricity price and resource requirements (copper for transformers and lines), peak load reduction is also possible
- ideal interface for the integration of future technologies (e.g., fast charging points for e-vehicles, forming of batteries for new vehicles, second life for electrical energy storage from the field)

This makes the DC grid particularly suitable for all tasks in factory automation. Typical applications are production lines, machining centers, robots, conveyors, hoists and test benches. A large number of variable rotation speed drives are used in these systems and machines. They generate significant amounts of regenerative power during braking.

From an energy point of view, the DC grid will be less suitable where individual base load drives are in continuous operation. Such applications exist, for example, in the process industry, in water supply or in large conveyor systems. Of course, this does not rule out the possibility that the use of a DC grid may also make sense here due to other advantages.

## 2.2 Partially Self-Sufficient Energy Supply Using Microgrids

The first attempts to establish electricity in society failed because it was provided by expensive electrochemical storage systems. Only the transmission of electrical power allowed the spatial separation of generation and consumption. Electricity developed into an energy transfer medium that is still only stored temporarily to a small extent [Ditt2013].

As the quota of processes in industry and everyday life supplied by electricity increased, the initial island grid structures transformed into interconnected grid. Today's interconnection of the largest possible number of consumers and producers results in both economic and technical advantages. According to Strauß [Stra2009], an interconnected grid can avoid the high costs that arise from storing sufficient amounts of electrical energy. The reason for this is that the fluctuating demand for electrical power is stochastically balanced out with the increasing number of consumers in the total load. As expected, the installed power plant capacity grows with the increasing size of the interconnected grid. However, the capacity, which is reserved for load fluctuations, decreases in proportion.

Looking at the development of the energy system from the perspective of the energy system turnaround – i.e., from the "Energiewende" –, the paradigm of interconnectedness and size is challenged by three aspects:

1. The capacity of plants making us of regenerative energy depends on meteorological conditions, so that conventional power plants, consumers and operators of energy storage facilities must adapt to the individual generative situation [PeekDiel2015], [Grass2015].
2. Compared to conventional generative plants, plants making us of regenerative energy are mainly connected to the distribution grid [Sola2018]. Therefore, new concepts have to be found to make the grid controllable at the transmission grid level and to integrate the decentralized generative capacities into the operational management of the grid [Dena2014].
3. If one compares regenerative and conventional plants, an essential difference is the lower average plant capacity of regenerative energy sources [BNA2018], [Lass2002]. Thus, the integration of renewable energy sources leads to the need of a more complex control system.

Figure 2.5 illustrates the effects of this development. While in the past the energy flow was mainly directed from the transmission grid to the distribution grid and finally to the end consumer (a), the integration of regenerative producers in today's grid state leads to uncontrolled and bidirectional balancing flows in the distribution grid (b). The concept for a DC grid presented in this book is based on this state-of-the-art. It presents local control zones as a solution for discussion (c). The literature often refers to these local control zones as microgrids.

The idea of a microgrid was first coined by the American scientist Lasseter [Lass2002] and has a wide range of applications in the literature. In the context of the DC grid concept presented, a microgrid is understood as a partially self-sufficiently operated electrical grid system, which integrates local consumers, decentralized generators and storage systems in one control area. The grid is called partially self-sufficient because it does not completely run without a connection to the external AC grid. It can be integrated into the superimposed grid control system as a control variable.

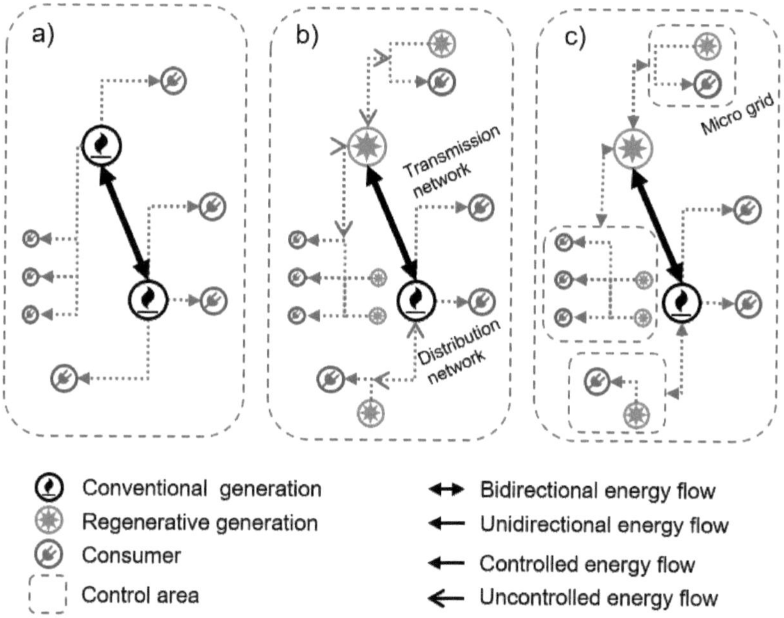


**Figure 2.5** Transformation of the transmission and distribution grid to a microgrid architecture

This microgrid architecture has macroscopic advantages for the entire electricity system because energy storage, generation and consumption within the local control zone are balanced in real time. This means that local balancing processes resulting from a change in the feed-in or consumption situation are not transferred to the superordinate grid. At the same time, the energy supply can be adapted from the superordinate grid structure by integration into its grid control system. The microgrid aggregates the connected load of all devices so that a relevant control variable for the next control hierarchy is created.

The local grid control opens up a spectrum of optimization options for operational management. Possible control objectives include, for example, grid failure safety, local consumption of regenerative energies, peak load capping or the energy-flexible operation of consumers.

## 2.3 Efficient Energy Supply

### Central Rectification/Merging of the DC Intermediate Circuits

*Omission of rectifiers in the devices*

In a DC grid, energy conversion is simplified. While today operating a drive with converter in an AC grid requires a rectifier, rectification is not necessary when used directly in a DC grid. If the device is to be capable of regenerating energy in an AC grid, an additional DC/AC converter must be integrated in the direction of the supply grid. This converter is also not required for direct DC supply.

The efficiency losses of the additional DC/AC converters in the direction of the supply grid are estimated to be 2.5% per conversion [Hamm2007]. Laudani has analyzed the efficiency of converters from different manufacturers and different power classes at different loads [Laud2015]. Converters with higher maximum power have a higher efficiency. Irrespective of this, the efficiency in partial load operation decreases. The analyses indicate a maximum overall efficiency of the inverters of only 92.5%.

*A better design of the central rectifier is possible*

A DC grid can achieve a further gain in efficiency through a centralized feed-in. The consistancy of the power demand in grids with an increasing number of participants leads to a better utilization of the central AC/DC converter at the supply point.

In contrast, partial load operation of less than 60% leads to additional efficiency losses of 2%.

*Use of recuperative energy*

In addition to the increase in efficiency, a DC grid, as already mentioned, enables the exchange of the regenerative energy (braking energy) of drives. While this energy is thermally converted in AC devices via braking resistors and is thus lost, a DC converter feeds this energy back into the DC grid with almost no loss. For an industrial robot in automobile production, Meike calculates a recuperation potential of 15% of the energy during operation and a total of 10% of the annual consumption, i.e., approximately 870 kWh [Meik2011].

If the electricity mix in Germany is taken as a basis, which produces 474 g $CO_2e$/kWh ($CO_2$ equivalent $CO_2e$ *Global Warming Potential*, indicated as $CO_2$-mass per kWh, which would produce the same greenhouse potential) [Umwe2019], this energy saving means an annual reduction of 412 kg $CO_2e$ per robot. In Germany about 27,900 new industrial robots will be used in 2018 [Litz2019]. Since these robots are of different sizes and purposes, a conservative estimate of the savings is based on the assumption that only half of the savings can be realized. The energy saving of 12.1 GWh would result in a $CO_2$ emission reduction of 5747 t $CO_2e$.

Economically, the energy savings through recuperation at an electricity price of 18.44 c/kWh including electricity tax means a cost reduction of 160 €/yr. per robot. Assuming a useful life of eight years for the robot, its energy costs over its life cycle are reduced by € 1280. With respect to the body-in-white production, for example, around 600 robots are used per model series. This means that energy costs can be reduced by € 768,000 over the life cycle, accelerating the amortization of a conversion to DC technology.

### No Capacitive Reloading Losses

DC lines do not have capacitive recharging losses as occur with AC lines. In the distribution grid this advantage is significant. In the case of motor lines, this advantage results indirectly from the fact that DC-fed motor converters are significantly smaller and thus these decentralized drives will be used more often. With these so-called integrated drives, the converter/inverter is directly connected to the motor, so the motor line is almost completely eliminated. This also eliminates the recharging losses in the motor line, which in the case of control cabinet inverters already generate significant losses when idle. These recharging losses, which are independent of the current motor current, are estimated at 30...70 W. Depending on the power of the drive, this corresponds to between 20% and 1% of the total rated power loss.

### Connection of Storage and PV

*DC/AC converters are replaced by more efficient DC/DC converters*

With values up to 99%, DC/DC converters have in principle (1–2 semiconductor switches, continuous current flow) significantly higher efficiencies than AC/DC or DC/AC inverters (6 semiconductor switches, intermittent current, commutating). Only with such energy-efficient converters the integration of storage becomes economically interesting. The same applies to the coupling with wind energy, photovoltaics and charging stations for e-vehicles.

When integrating storage systems, one can assume an increase in efficiency of 5–7.5% (elimination of one conversion stage each for charging and discharging). A PV system increases its efficiency by 2–3% owing to the elimination of the last DC/AC conversion stage.

*Storage management becomes more efficient*

For storage components that are operated on a DC grid with a wide working voltage range, a very simple and efficient form of storage management is now possible (see also Chapter 6.5). Since the level of the applied DC voltage provides information about the current load state of the DC grid, the charging and discharging phases of the storage can be optimized even without fieldbus and other measuring points in the system.

### Use Case Example

In one use case comparative measurements were carried out on a CNC machining center for woodworking. In order to get the best possible basis for a comparison between AC and DC variants, the existing AC axles were converted or modified so that they can be operated in a DC grid (see Chapter 9.1).

The direct comparison between the AC and DC versions of the CNC machining center results in an overall increase in efficiency of at least 5.5% for the same machining profile. This is achieved by recuperating the energy of the milling spindle and the main movement axles.

Figure 2.6 shows that the graph depicting the power of the DC version also assumes negative values. The peaks of down to −8 kW occur when the portal system of the machining center brakes before reaching a new target position, thus making the braking energy from the DC sector (grid of several DC units, see also Chapter 5.1) available to the DC distribution grid.

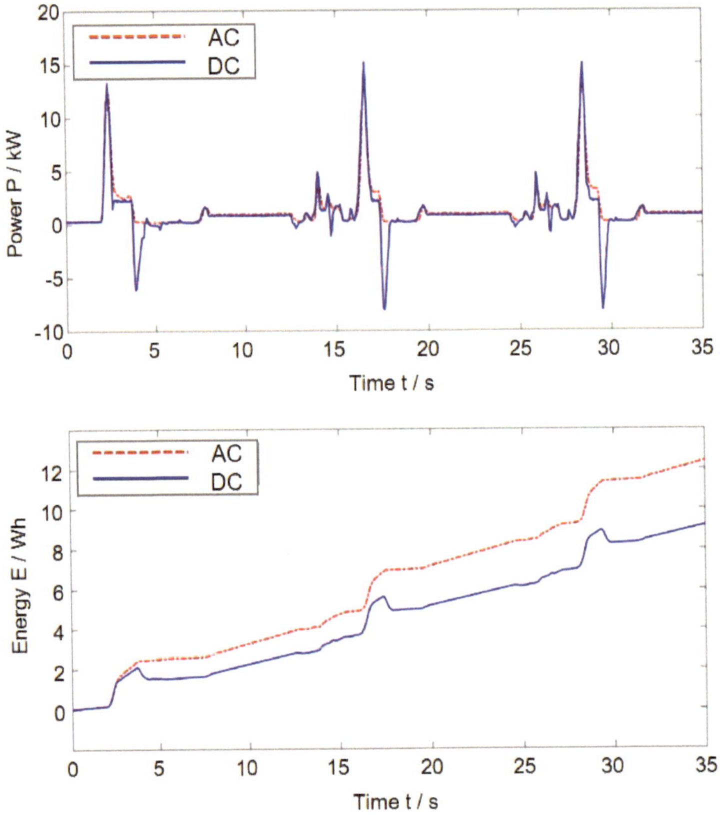


**Figure 2.6** Energy consumption of the DC sector "CNC feed drives" before and after conversion to DC

The savings through recuperation and through direct energy exchange between all drives in the shared DC grid amount to 2.2 Wh per cycle of 12 seconds and thus 0.662 kWh per machining hour. If five processing hours per day are assumed, this results in savings of 760 kWh/yr. or 360 kg $CO_2$e/yr. for an annual mileage of 230 days. In terms of energy costs, such a reduction means a cost reduction of 140 €/yr.

### Total Potential

Overall, in an industrial context the DC grid has the potential to reduce the energy required by about 10–12%. The savings potential is increased by the following factors:

- Masses that are accelerated and decelerated and their dynamics. Dynamic processes such as robotic applications or rapid machining with changes of direction have a correspondingly high potential for recuperation.
- Degree of energetical griding in the existing application, i.e., total size of the DC grid. The more components and devices are grided with each other and the larger the DC grid is, the lower the simultaneity factor is. Short-term energy requirements and surpluses can be efficiently compensated. In an extensive DC grid, the distributed capacity, i.e., the electrical power that can be stored in capacitors, also increases. With the grid increasing, larger individual peak loads can be absorbed.
- Integration of local energy generators (e.g., photovoltaic systems, combined heat and power plants) and/or electrical storage for short-term buffering or even holding back energy. Especially for small systems or machines that require or generate short-term power peaks, a storage system can be a useful extension.

*New components that are necessary for operating a DC grid*

The gain in efficiency is offset by the additional energy required to control new types of protective elements. In the case of an electronic circuit breaker, additional losses occur in the power path, since all use energy must be conducted via semiconductors. Since there are no commercially available devices to date, no meaningful value can be given. The devices that were developed in the DC-INDUSTRIE project are functional models. Most of these devices were built discretely and are therefore energetically not comparable with possible series products. However, one can assume that no significantly higher losses occur here than do regarding those energy measuring devices, that are already in use today for intelligent energy management in an AC grid. Measuring devices of this kind are not required when implementing a DC grid because they can be functionally integrated into the protective components.

# ■ 2.4 Resource-Saving Energy Supply

### Rectifiers Are Omitted

The omission of a conversion stage in devices such as frequency converters or PV-inverters results in savings in semiconductor components, electronic components and printed circuit board real estate. Since this saving of resources has not been calculated on the level of individual components so far, an estimation of the value is made at this point. The market research carried out in DC-INDUSTRIE of publicly known list prices of different manufacturers has resulted in a reference price of 15 €/kW for the cost saving of DC devices.

Using this value together with the key figures of the electronics industry, it is possible to estimate a $CO_2$ footprint in $CO_2e$ equivalents (relative greenhouse gas potential compared to the greenhouse gas potential of the same mass of $CO_2$). The Green Design Institute of Carnegie Mellon University has developed an online tool for this purpose [Gree2019]. The values are taken from a 2007 benchmark of the U.S. industry. The estimation took the following sectors into account: semiconductors, integrated circuits and electronic devices, capacitors, resistors, transformers and connectors, relays and industrial controls (Table 2.1). In order to further detail the industrial sectors, the relevant ones were selected and included in the evaluation as an average value. The conversion from US Dollar to Euro is based on an exchange rate of 1.35 USD per Euro.

**Table 2.1** $CO_2e$ of Different Electronics Industries in Relation to Sales

| Sector | t $CO_2e$/million $ | kg $CO_2e$/$ |
|---|---|---|
| Semiconductors | 429.3 | 0.429 |
| Integrated circuits and electronic devices | 391.5 | 0.392 |
| Capacitors, resistors, transformers and connectors | 401 | 0.401 |
| Relays and industrial controls | 334.8 | 0.335 |
| Average | 389.1 | 0.389 |

Assuming that costs of 15 € per kW per device can be dropped, 5.835 kg $CO_2e$/kW could be saved. With a total drive power of 70 kW in a CNC machining center, 408 kg $CO_2e$ could be saved in the converters. The supplying active infeed converter (AIC) of the same machine would have to provide only 30 kW due to a simultaneity factor significantly less than 1 and the use of the recuperated braking energy. Such an AIC causes an estimated 175 kg $CO_2e$, which reduces the above-mentioned savings. In total, a saving of 233 kg $CO_2e$ per machine would remain. In 2010 german companies produced about three million variable-speed drives in the power class 0.75 kW to 7.5 kW [Wiet2010]. Assuming that 30% of these were speed-controlled, about 900,000 converters are needed. It can be assumed that smaller drives are increasingly used, so that an average power of 1.5 kW is assumed. This would save 1.35 GW of rectifier power in the devices. In terms of the $CO_2$ footprint, this would represent a potential of just under 7900 t $CO_2e$/a.

### Capacitors and Assembly Space Become Smaller

DC link capacitors are central elements in classic AC converters. Comparatively they require a lot of installation space and cause high costs. A well-planned dimensioning should take several factors into account. The capacitors must be able to buffer enough energy to suppress impermissibly high voltage ripples from the output inverter. At the same time, they must be able to cope with peaks in the recharging current that are caused by the input rectifier (see also Chapter 4.3, last paragraph).

With DC inverters, the requirements regarding the capacitors change. Omitting the rectifier - and thus the opening/expansion of the DC link - has two decisive advantages, which in principle favor a smaller design of the capacitors:

1. The capacitors do not have to buffer all the energy that is required between two recharging phases of an AC inverter. Due to the distributed capacity and the smaller simultaneity factors, maximum and peak loads can be drawn from the immediate vicinity of the DC grid without the voltage at the capacitors dropping too much.
2. There are no high recharging currents through rectifiers anymore. The central rectification eliminates all effects of diode rectifiers in the DC converters. The current flow is more even and largely dependent on the load and not on the state of charge of the capacitors.

Further studies in the context of the project have shown that DC inverters generally require much less capacity per kW rated power in order to not impair the stability of the DC grid and the service life of the devices (see also Chapter 4.3, last paragraph).

Taking into account different device sizes, type series, local multi-axle grids and capacitor technologies, estimates show that the capacity required for DC inverters can be reduced to ⅓ on average.

### Lower Installed Load

Owing to the reduction of the total energy demand and the reduction of peak loads, considerably lower potential peak power levels are required than would be the case with classic AC distribution grids. As a result, the transformer stages for the low-voltage grid can be dimensioned smaller. The same applies to the associated lines and filters. Copper and iron can be saved.

### Less Cable Material

A further potential for saving resources can be found in the cables. While an industrial AC grid uses a 3-phase grid with a protective conductor and designed with or without a neutral conductor, the DC grid is designed as a 2-wire grid with a protective conductor. This eliminates the need for one or two conductors in the cabling.

The difference in material usage can be illustrated using a 2.5 $mm^2$ cable. In the case of a 3-wire cable, the weight of copper is 72 kg/km. The entire total cable weighs 132 kg/km (60 kg/km for insulation material). A four (five) core cable has a copper content of 96 (120) kg/km with a total weight of 163 (200) kg/km. The weight of the insulation material is 67 (80) kg/km. The DC grid saves 25% (40%) copper and 10% (25%) insulation material while allowing for the same transmittable power. The copper allows savings of 692 kg $CO_2$e/t for recycled copper and 5152 kg $CO_2$e/t for primary copper [Fris2010]. Taking PVC as insulation material would allow savings of a further 230 kg $CO_2$e/t for recycled granules and 2100 kg $CO_2$e/t for new material. For the DC grid, this leads to savings of 18 kg $CO_2$e/km in relation to a four-core AC grid or 38 kg $CO_2$e/km with a five-core AC grid.

## 2.5 Robust and Flexible Power Supply

The technically simple integration of storage systems into industrial DC supply grids opens up promising possibilities for storage utilization. Different use cases can be supported:

- Power reduction/load peak reduction
- Load shifting with economic optimization
- Increase in reliability

The reduction of power and load peaks for the electrical supply from the AC grid is achieved by an adapted system of different storage systems. These can be arranged in direct proximity to the process causing the power or at a higher grid level.

The DC grid in its developed form can act as a DC Smart Grid, dividing the power flow between different storage devices. In a process with low dynamics, a battery can both reduce the connected load and ensure uninterrupted operation or a guided shutdown of the system (Figure 2.7). Higher dynamics require different storage technologies. A capacitor, for example, compensates for short-term power boost. If more energy is required, a flywheel storage unit is a useful addition to the capacitor. The interaction of these storage technologies has been successfully implemented and tested in initial test facilities.

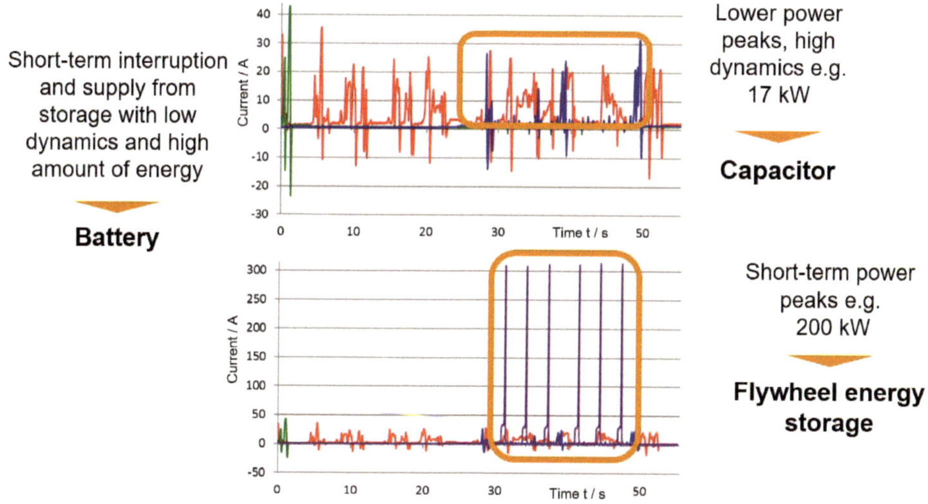


**Figure 2.7** Applications of different storage technologies

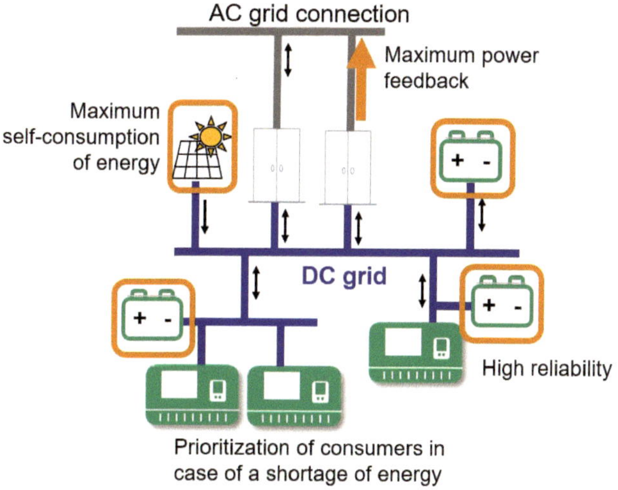


**Figure 2.8** Optimization of the energy supply system in relation to economic and ecological goals

If storage facilities are integrated into an energy supply system, it can be optimized in relation to varying energy tariffs. In such a case, the system changes the operating points of the storage units by means of a superordinate grid control, so that they provide more or less energy. For example, the extent to which self-generated energy is used can be influenced in this way (Figure 2.8).

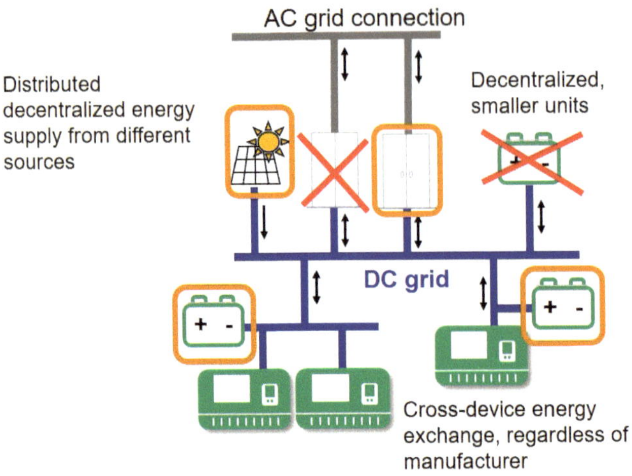


**Figure 2.9** Optimization of the energy supply system in case of supply bottlenecks

If the supply fails, e.g., due to disturbances in the supply grid or due to technical defects, storages and also self-generation systems can immediately take over the energy supply (Figure 2.9). By changing the voltage level in the DC grid, each participant is informed about the energy shortage in real time. Flexible processes may independently switch off or reduce their energy demand thereby additionally stabilizing the grid.

In 2017 the Bavarian Chamber of Industry and Commerce (BIHK) published a study on the quality of supply [Bihk2017]. The study classifies the ratios of short-term interruptions over their duration (Figure 2.10).

It becomes clear that interruptions of less than one second – statistically not yet recorded – occur significantly more frequently than unplanned supply interruptions of more than three minutes – which are regularly reported. If one uses a ratio of 10:1 one can extrapolate from the 52,000 unplanned supply interruptions >3 min as determined by the Federal Grid Agency in 2017 that the number of short-term interruptions of <1s might be around 520,000 instances. A number of 869 low-voltage grids is cited as a reference value [Bund2018]. According to this, there are just under 600 interruptions per year per grid and thus about 1.6 interruptions per day.

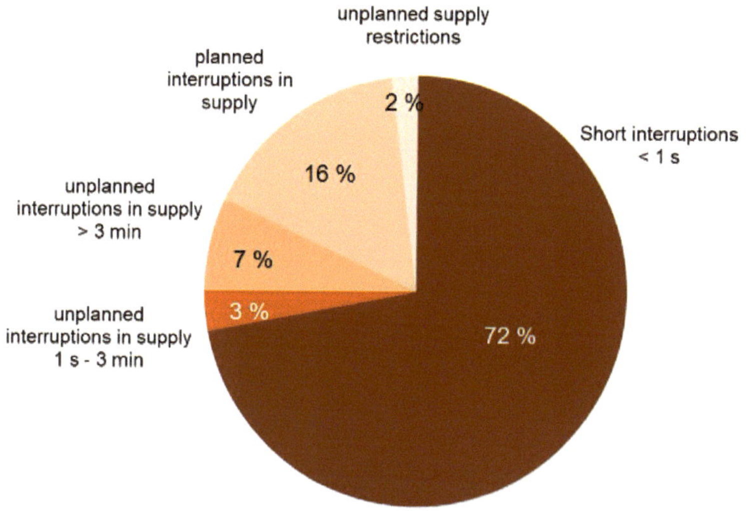


**Figure 2.10** Breakdown of supply disruptions by type of incident [BIHK 2017]

The disruptions in the supply grid cause high investments and additional expenses for individual companies [Bihk2017]:

- “An uninterrupted power supply (UPS) based on lead accumulators - which can only protect one of our printing machines - generates power losses that result in additional costs of over 50,000 Euro/year in the procurement of electricity.” Markus Fuchshuber, Franz Anton, Niedermayr Graphische Kunstanstalt GmbH & Co. KG
- “The possibility of energy shortages due to little wind or solar power generation without sufficient conventional generation in the background has prompted us to take double precautions by using emergency power generators.” Johannes Bock, HEINZ Glas GmbH
- “Due to the influence of supply quality problems on production, we have invested in more intensive monitoring after the various process steps. In our fully automated processes, this is the only way we can prevent defective products from being further processed.” Herbert Klein, Agfa-Gevaert HealthCare GmbH
- “In 2013/14, we had more than ten identified outages over the course of 14 months, which we attribute to poor supply quality. Since then, however, the situation has improved.” Herbert Klein, Agfa-Gevaert HealthCare GmbH
- “In 2016, we had five machine downtimes due to undervoltage, each of which caused an operating loss of 2 to 15 hours.” Markus Fuchshuber, Franz Anton, Niedermayr Graphische Kunstanstalt GmbH & Co. KG

- “Even a standstill of one minute causes a production loss of about one hour.” Johannes Bock, HEINZ Glas GmbH
- “In our company, the time required to resume operations after a short interruption ranges from a few minutes to longer periods of time.” Thomas Küfner, Brose Fahrzeugteile GmbH & Co. KG
- “We are currently seeing two developments that are in contrast to each other. On the one hand, the voltage quality is getting worse due to non-linear loads such as converters. On the other hand, new devices are becoming increasingly sensitive to deviations from the perfect sine wave.” Walter Huber, Stadtwerke Bad Tölz GmbH
- “We have noticed that customers' systems are becoming increasingly sensitive to short-term voltage dips. Installed systems often do not comply with the FNN's recommendations on immunity. The grid operators urgently recommended that systems should be able to handle voltage dips of up to at least 100 milliseconds. In reality, however, this only occurs at much shorter times.” Johannes Brantl, Bayernwerk

The DC feed increases the system's resistance to interference from the AC grid. In the event of voltage dips or faults in the AC grid, these are not passed on directly to the DC grid, as the feeder prevents a power flow from the DC grid into the AC fault. The distributed capacities built into the grid lead to an intrinsic robustness. They are able to retain energy in case of short-term supply interruptions.

This was calculated exemplarily for the CNC machining center mentioned above. In this machine, devices with a total nominal power of 100 kW were installed (without storage connection). The total capacity of the devices in the DC grid had a value of approx. 2.5 mF. Considering an average power requirement of 10 kW and depending on the grid voltage, this results in grid support times for different capacitors as listed in Table 2.2.

**Table 2.2** Backup Support Times through Internal Capacitors

| | Voltage | Average grid support times |
|---|---|---|
| Nominal voltage | 650 V | 7.8 ms |
| Minimum | 500 V | 1.3 ms |
| Maximum | 750 V | 15.3 ms |

Voltage dips of 7.8 ms (almost a half-wave in the AC grid) can therefore normally be bridged without a complete shutdown of the system. No additional costs arise in this case. Beyond these 7.8 ms, additional storage units must be integrated. In larger grids, this value increases due to the lower average load power in relation to the sum of the devices’ nominal power.

# 2.6 Economic Perspective

The technological comparison of the AC grid with the DC grid as well as the efficiency evaluations have shown the essential differences and thus the cost influencing factors:

- Centralization of the feed-in
- Omission of rectifiers in frequency and servo converters
- Reduction of the number of conductors in lines
- Necessity of new protection components with power semiconductors

Within the research project DC-INDUSTRIE two model plants were realized firstly in AC-technology and then in DC-technology. The first plant has two handling systems and a connecting transport system with roller conveyor (intralogistics system). The second is a CNC machine converted for woodworking (machine tool).

In order to determine the cost differences between the realization of the two test plants in first AC and then DC the differences in equipment cost were examined. For this purpose, a market research determined standardized device prices for the supplying Active Infeed Converter (AIC) and for the converters of the speed-controlled drives. For the AIC, a linear relationship between device costs and device performance can be specified (Equation 2.1).

$$C_{AIC} = C_{fix} + C_{var} \tag{2.1}$$

The fixed costs $C_{fix}$ cover basic features of the AIC, such as the casing, the connections, and the communication and control unit. The variable cost component $C_{var}$ depends on the performance. These costs include expenses for the power semiconductors and for cooling.

The device costs of the converters for the controlled drives depend on the power, the installation location and the functionality. Cost curves with fixed and variable costs for both AC and DC devices can be determined. The costs for the protection components - which are not yet available - are estimated using functionally similar device combinations in the AC grid.

Using these estimates of the costs for the individual devices, the changes resulting from the conversion of the systems to DC supply are evaluated.

Figure 2.11 illustrates the cost differences in relation to the total costs for AC supply and AC drive technology without motors and controllers. Regarding drive technology, the costs for the converters and their supply units were considered. Regarding grid system technology, all component costs for filtering, measuring devices and the 24 V supply are included. The line costs also include the costs for the connectors. The cost group "protection technology" includes the costs for the fuse components. AC grid connection, dominated by the costs for a transformer, and DC supply technology are costs that only occur in a DC grid.

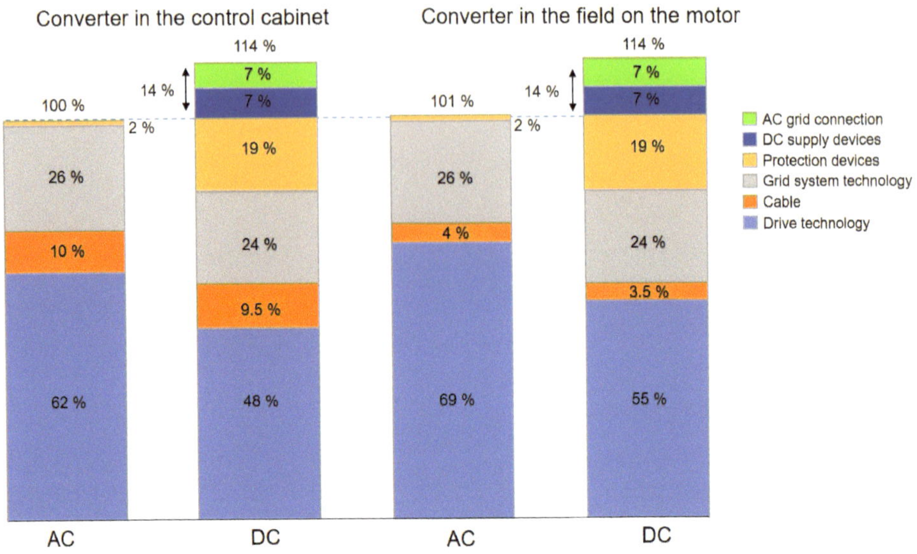


**Figure 2.11** Cost comparison AC and DC, based on components of the energy supply in a pilot plant (intralogistics system)

It can be seen that direct drive technology with converters for the speed-controlled drives as well as their supply is becoming cheaper. In the example, the AC system is based on the assumption that the servo axles for the two handling systems are connected in a DC link circuit. Although the costs for the actual lines are lower due to the reduction of the wires, the total line costs are almost the same. This is particularly due to more complex plug connections and other insulation materials. The costs for the protection technology in the DC realization increase significantly. The reason for this boost in costs lies in the DC circuit breakers (in this project they are called Smart Breakers), which provide other functions besides the actual protection, such as energy measurement and precharging of the DC grid.

Overall, the cost difference of the components for the supply and drive technology is 14%. About half of this is attributable to the isolating transformer (AC grid connection). When supplying an entire workshop, the transformer costs are eliminated, since conversion to DC can take place directly at the workshop's supply point. To take advantage of the scale effect of the DC system, cost equality can be achieved when supplying about eight handling systems and four times the size of the transport system. The scale effect is particularly favored by the fixed costs of the AIC. It is also supported by the reduction in power of the central AIC compared to the sum of the power of the individual supply modules. In addition, the costs for the protection technology for the DC supply are reduced for the same reason. It is noticeable that when changing to a partial installation of the converters directly on the motor (field devices), the device costs remain virtually constant. The additional costs for the inverters (higher protection class of the casings, changed cooling concept) are compensated by the savings in motor cables. Not included in the costs are the expenses for cable laying and cable routing (e.g., energy chains).

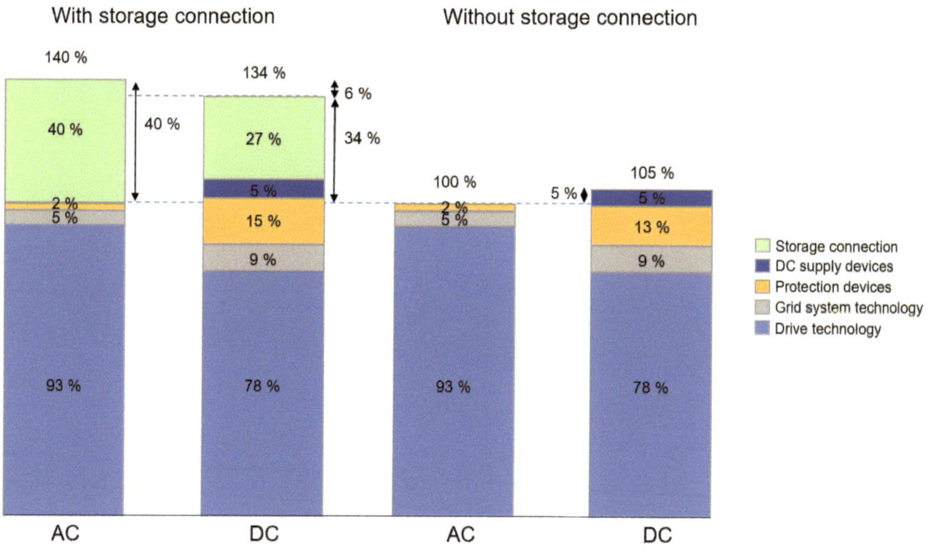


**Figure 2.12** Component cost comparison AC and DC in a test plant with storage (machine tool)

The cost comparison for the CNC machining center shows a slightly different picture (Figure 2.12). Here, more complex individual converters were used for the positioning and control of the drives. For process engineering reasons, the system deliberately does not make use of a DC link system. In the DC grid, the DC converters in the individual devices are omitted, so that a significant cost reduction of 15% in relation to the total costs can be achieved. This saving partially compensates the additional costs for the protection technology (+11%), the grid system technology (+4%) and the central DC supply (+5%). A transformer was not required due to the simple central DC supply unit. All in all, the DC grid leads to a cost increase of 5% in the machine tool for the electrical drive system without the motors and cabling.

A special feature can be seen in the left part of Figure 2.12. The use of a storage system for power reduction can be implemented much more cost-effectively in the DC grid. The savings amount to about 13% due to the omission of bidirectional inverters and the control system for regulating the storage system on the grid.

The economic potential of DC grids depends on the following factors:

- Larger systems make use of the economies of scale of a better supply of the inverters.
- The integration of storage systems can be carried out more cost-effectively.
- Today's use of a local DC link is a good basis for expanding towards an even more cost-efficient DC grid.
- The protection technology currently causes the greatest additional costs in a DC grid, as there are no widely used components available on the market.
- The separate isolating transformer can be omitted for large systems that are directly connected to the workshop supply.

## 2.7 Summary

DC technology offers various potentials to make production more efficient, more reliable and therefore more economical. There are as many implementation possibilities as there are applications. In order to decide where the use of DC technology makes sense and which advantages can be used, the following considerations may help.

**Primary Economic Potentials:**

If a plant or factory has a lot of dynamic kinetic energy, the proportion of regenerative energy is usually quite high. In a DC grid, this energy can be reused, whereas in an AC grid it is usually lost at a braking resistor. Therefore, in a DC grid energy costs for production can be saved.

Due to the possible reduction and equalization of the total energy demand, transformers and lines can be designed for lower peak loads, which saves material and thus costs.

**Secondary Economic Potentials:**

Due to its open architecture and distributed capacities, a DC grid already offers a natural possibility to absorb short grid interruptions. The AC grid does not offer this feature, which means that even grid errors lasting less than one second can lead to production downtimes. For longer grid failures and also total failures, the DC grid offers in principle simpler possibilities to maintain production by integrating storage and decentralized energy generators. This reduces production interruptions, especially in regions where a stable energy supply cannot be taken for granted. Particularly in highly automated production facilities such as in the automotive industry, long production downtimes as a result of power failures cause great economic damage.

**Tertiary Economic Potentials:**

In addition to saving energy and resources while increasing availability, a DC grid generally offers the possibility of integrating regenerative energy sources more easily. It thus promotes decentralized energy generation. Together with a local base load supply, the dependency on energy supply companies can be partially or even completely eliminated. In areas where island grids are unavoidable, DC technology is already a particularly economical alternative today. This applies, for example, to distribution grids on ships that are primarily fed by the main diesel engine.

## Literature

[Hamm2007] Hammerstrom, Donald: AC Versus DC Distribution Systems – Did We Get it Right? Proceedings of IEEE Power Engineering Society General Meeting. Institute of Electrical and Electronics Engineers, Tampa, 2007

[Laud2015] Laudani, Giuseppe: Comparison of cost and efficiency of DC versus AC in office buildings. Imperial College, London, 2015

[Meik2011] Meike, Davis; Ribickis, Leonids: Recuperated Energy Savings Potential and Approaches Robotics. IEEE International Conference on Automation Science and Engineering. Institute of Electrical and Electronics Engineers, Trieste, 2011

[Umwe2019] Umweltbundesamt: Entwicklung des $CO_2$-Emissionsfaktors für den Strommix in Deutschland in den Jahren 1990 bis 2018. Umweltbundesamt, 2019, https://de.statista.com/statistik/daten/studie/38897/umfrage/co2-emissionsfaktor-fuer-den-strommix-in-deutschland-seit-1990/ (access on 24 Apr 2020)

[Litz2019] Litzenberger, Gudrun; Dr. Bieller, Susanne; Dr. Müller, Christopher: World Robotics 2019 Preview. IFR International Federation of Robotics, Frankfurt Main, 2019

[Gree2019] Green Design Institute: Economic Input-Output Life Cycle Assessment. Carnegie Mellon University Green Design Institute, Pittsburgh, 2019, http://www.eiolca.net/ (access on 02 Sep 2019)

[Wiet2010] Wietschel, Martin; Arens, Marlene; Dötsch, Christian; Herkel, Sebastian; Krewitt, Wolfram; Markewitz, Peter; Möst, Dominik; Scheufen, Martin: Energietechnologien 2050 – Schwerpunkte für Forschung und Entwicklung. Fraunhofer-Verlag, Stuttgart, 2010

[Fris2010] Frischenschlager, Helmut; Karigl, Brigitte; Lampert, Christoph; Pölz, Werner; Schindler, Ilse; Tesar, Maria; Wiesenberger, Herbert; Winter, Brigitte: Klimarelevanz ausgewählter Recycling-Prozesse in Österreich. Federal Environmental Agency, Vienna, 2010

[Bihk2017] BIHK-Studie: Energiewende im Strommarkt – Versorgungsqualität. Industrie- und Handelskammer in Bayern, München, 2017

[Bund2018] Bundesnetzagentur: Kennzahlen der Versorgungsunterbrechungen Strom. National Regulatory Agency of Germany, Bonn, 2018

[Ditt2013] Dittmann, Frank: Rückblick auf die Anfänge der Elektrifizierung. Springer-Verlag, Berlin, 2013

[Stra2009] Strauß, Karl: Kraftwerkstechnik. Zur Nutzung fossiler, nuklearer und regenerativer Energiequellen. 6. Auflage, Springer-Verlag, Berlin, 2009

[Peek2015] Peek, Markus; Diels, Robert: Strommarktdesign der Zukunft. Umweltbundesamt, Dessau-Roßlau, 2015

[Graß2015] Graßl, Markus: Bewertung der Energieflexibilität in der Produktion. Herbert Utz Verlag, München, 2015

[Dena2014] dena-Studie Systemdienstleistungen 2030: Sicherheit und Zuverlässigkeit einer Stromversorgung mit hohem Anteil erneuerbarer Energien. Deutsche Energie-Agentur GmbH (dena), Berlin, 2014, https://www.dena.de/fileadmin/dena/Dokumente/Pdf/9094_dena-Studie_System dienstleistungen_2030.pdf (access on 24 2018)

[Sola2018] Bundesverband Solarwirtschaft e.V.: Statistische Zahlen der deutschen Solarstrombranche (Photovoltaik). Bundesverband Solarwirtschaft e.V., Berlin, 2018, *https://www.solarwirtschaft.de/fileadmin/user_upload/bsw_faktenblatt_pv_4018_4.pdf* (access on 19 Nov 2018)

[BNA2018] Bundesnetzagentur für Elektrizität, Gas, Telekommunikation, Post und Eisenbahnen: Kraftwerksliste Bundesnetzagentur (bundesweit; alle Netz- und Umspannebenen) Stand 2.02.2018. *https://www.bundesnetzagentur.de/SharedDocs/Downloads/DE/Sachgebiete/Energie/Unternehmen_Institutionen/Versorgungssicherheit/Erzeugungskapazitaeten/Kraftwerksliste/Kraftwerksliste_2018_2.xlsx?__blob=publicationFile&v=3* (access on 19 Nov 2018)

# 3 Requirements for Industrial Energy Supply

Timm Kuhlmann
Herbert Graf
Patrick Kröhner
Tobias Laubscher

Demands on future industrial energy supply are manifold. They are determined internally by production and its processes. The increasing demands on energy supply quality and efficiency derived from this are in conflict with the changing public energy supply grid. In relation to this, demands are also being placed on the industrial energy supply, for example in terms of temporarily reducing its use of power. Tomorrow's factories will have to master the balancing act between a volatile public energy supply with increasing disruptions and higher demands for flexibility and the process-related requirements for quality and efficiency. In order to create clarity here for the development of a DC-powered factory, different manufacturing companies were investigated with regard to their requirements.

The relevant requirements were initially identified by interviewing experts from the fields of electrical system planning and supply planning in three selected manufacturing companies. The requirements were then specified and clustered.

## 3.1 Requirements from the User's Point of View

The expert survey revealed a total of 92 different requirements for a DC-based energy supply in production. They were clustered into seven areas thereby reflecting the thematic priorities in the development of a direct current grid. As part of the detailing process the clusters were weighted. A distinction was made between mandatory, important and optional requirements (Figure 3.1). If not fulfilled, mandatory requirements would prevent the DC grid from being used in production. Important requirements should be implemented from the user's perspective but were not considered mandatory for initial applications. However, broad success is questionable if these important requirements are not covered by solutions. Optional requirements could be implemented or not. They would, if implemented, offer a respective added value from the user's point of view.

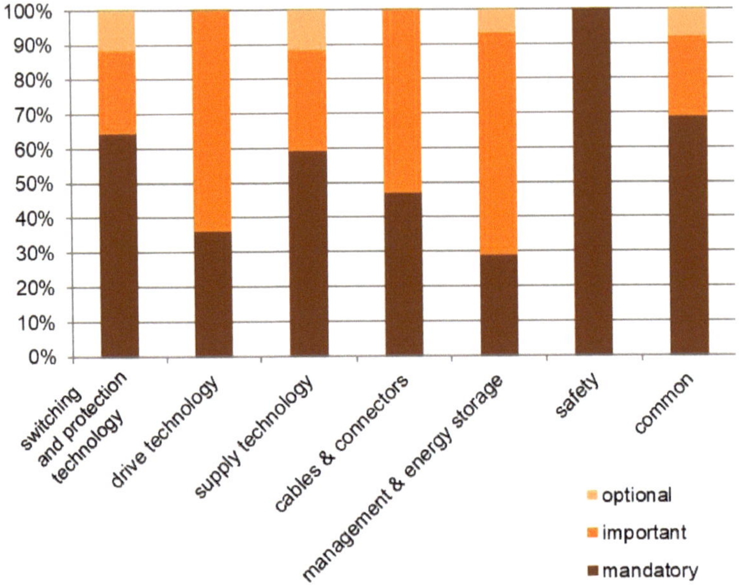


**Figure 3.1** Categorization of the requirements for a DC grid

Figure 3.1 shows the areas on which development is focused. Most of the mission-critical requirements address switching and protection technology. Together with the safety of the grid, there is little scope here for weakening the requirements. Supply technology is also characterized by a high proportion of mission-critical requirements. Robustness and efficiency are the two main requirements. Cables and connectors are a further yet different group. Compatibility as well as safety and protection when disconnecting under load are important requirements here. General requirements include those for environmental conditions, service life, expansion, and compliance with norms and standards. Grid management and integration of energy storage are important requirements. In summary, the DC grid should operate robustly and be as easy to handle as any previous AC grid. Last but not least there is the area of drive technology. Here the objective is to integrate decentralized drives in particular in an optimized way into a factory's internal distribution grid. Precise knowledge of the requirements in the individual areas makes it possible to develop the DC grid in a highly application-oriented way and thus to efficiently exploit the potential identified in Chapter 1 and Chapter 2.

### Requirements for Switching and Protection Technology

The requirements for switching and protection technology are derived from practical situations in the industrial environment. Three situations can be distinguished: maintenance, emergency stop and faults in the DC grid.

In order to maintain machine parts, a safe disconnection via a maintenance switch must be available. The respective grid section must be disconnected, and an energy feed prevented. Capacities that are still charged must be discharged or disconnected from the mains in the same way as energy sources. The switched-off machine section stops. This may generate regenerative energy from the decelerating drives. This energy must either be fed into the storage systems before disconnecting, or it must be discharged in a similar way to the capacitors. The top priority at this point is to prevent any flow of energy within the switched-off machine section as well as an energy flow into the respective machine section. If the energy flow has dropped to zero and the voltage has been switched off, the standards according to DIN VDE 0105 of the *Berufsgenossenschaft*[1] must be applied. This means that, especially in the DC grid, short-circuiting must be possible after checking that the voltage has been disconnected without damaging any equipment or restricting the function of the higher-level DC grid. Such disconnection for maintenance purposes must be possible in principle for every part of the machines. This can be the entire machine or smaller subsystems, right down to a single drive.

DIN EN 60204-1 VDE 0113-1:2019-06 differentiates between emergency halting (German: Not-Halt) and emergency stopping (German: Not-Aus). While emergency halting is intended to particularly protect in emergency situations caused by hazards emanating from drives, the emergency stop function focuses on hazards emanating from electrical energy. DIN EN ISO 13850:2016-05 “Sicherheit von Maschinen - Not-Halt - Gestaltungsleitsätze”[2] stipulates a stop category 0 and a stop category 1 regarding emergency halting. Pertaining to electrical energy flow, the difference in the stop categories is the interruption of the electrical energy supply. Stop category 0 requires an immediate mechanical interruption *without creating other hazards*. This immediate interruption is not required in stop category 1. Here, a control system that stops the hazardous movement as quickly as possible is initially sufficient. The energy supply is only interrupted after the movement has come to a standstill and thus the danger has been averted. There is no obligation to disconnect the energy supply electrically. The standard expressly points out that no further hazards may be generated during disconnection. This also defines the wording “in the fasted possible way”[3] or “as fast as possible”[4]. If - owing to sys-

---

[1] In Germany the *Berufsgenossenschaft* is an employer's liability insurance association.

[2] “Safety of Machinery - Emergency Halting - Design principles”

[3] German: “schnellstmöglich”.

[4] German: “so schnell wie möglich”.

temic reasons – delays occur, a separate hazard assessment must be carried out taking into account all the technical possibilities of energy reduction in the case of an emergency halting. The hazard can hence be seen as an effect of the original situation or of the action of disconnecting as such.

An emergency stop (German: Not-Aus) is intended to protect against electrical energy in emergency situations. This function is to be provided where protection against direct contact can only be ensured by means of a barrier or of obstacles and thus as a safety distance. Examples are slip ring systems or switchgear in electrical operating rooms. In addition, an emergency stop must be provided where there is a possibility of other hazards or damage from electrical energy. In this context, the standard also refers by way of example to the danger of stored energy being released, which can lead to an electric shock or an unexpected movement of the machine. The energy can be stored electrically in capacitors or batteries, or mechanically in a movable mass, and fed into the grid through recuperation. In addition, after a supply is switched off, the residual voltage must either fall below 60 V within five seconds or appropriate warnings must be given to indicate an extended discharge time.

The objective must be that the system itself is capable of protecting against such hazards. This can be achieved through a system and equipment design that is aware of the risks and hazardous situations and reduces them to a minimum through appropriate technical measures. The remaining risks must be protected by training, by structural measures on site and by emergency stop systems.

Short circuits and ground faults can occur in the DC grid; limiting the resulting damage is the task of protection technology. Short circuits are detected by overcurrent protection devices such as fuses. If a short circuit is detected, it is the task of the fuse to stop the current flow as quickly as possible. Today's solutions differ, among other things, in their capability to reclose. While a fuse cannot be reused after being triggered in case of a short circuit, circuit breakers can be reclosed after a fault. Fusible links should be avoided for a quick restoration of normal grid operation, as this requires additional effort for stocking and providing the fuse. Users favor reclosers as protection devices. Initial implementations have shown that smart protection devices offer certain advantages. These will be described in detail in the course of the book. However, significantly higher costs must be expected than for fuses. Therefore, a protection concept for DC grids is necessary that on the one side ensures safety while on the other gets by with the smallest possible number of protective devices. With regard to a residual current device (RCD), which detects current flows via the protective conductor and thus the grounding, the requirements are divided. Some industries do not use RCDs today and see no need for them in the future. In plants outside Germany and in certain plant concepts, an RCD is mandatory or is currently installed by the plant manufacturers. The components of the DC grid must be suitably designed for these applications.

In general, switching and protection devices must have a life expectancy comparable to today's solutions in the AC grid. In addition, it must be noted that semiconductor components connected within the continuous power flow cause energy losses. These losses must be taken into account in the overall consideration.

### Requirements for Drive Technology

About 70% of the electrical energy within production is required for drive technology. Drive technology thus plays a decisive role in the implementation of DC grids. It is the power ranges between 0.37 and 7.5 kW that dominate drive technology, and that will continue to increase in the future due to flexibilization and automation. Even though the tightening of regulations concerning the efficiency of electric motors is increasing the focus on frequency-controlled drives, unregulated fixed-speed motors operated directly from the AC grid are still used in intralogistics, for example. Such cost-effective drive solutions easily be implemented in a DC grid.

From the perspective of the supply grid, it is the frequency inverter that plays a special role in variable-speed drive solutions, as it forms the interface between the motor and the grid. Different installation locations are possible for the frequency inverter. Control cabinet devices are conceivable, but also devices that are installed on the outside of the safety fence as well as directly on the motor. This results in further requirements, for example regarding cables or plug connections. Cooling concepts change due to the protection class of the casing, but also due to the internal structure of the devices in the case of a direct DC supply. According to the safety concept, solutions for emergency stop or failure of the DC grid must be provided, which can handle any energy or voltage peak due to recuperated mechanical energy at the drive. In addition, the converters for the motors must be capable of handling overloads. Common requirements in this context are an overload capability of 200% for three seconds or 150% for 30 seconds.

The main task of a direct DC supply of a drive system, that consists of an inverter and motor, is to increase energy efficiency. The requirement is to achieve the highest international guidelines, i.e., ones that stipulate compliance with energy efficiency classes of the drives according to IE classes.

Additional requirements do not directly concern the DC supply of the inverters but relate to their range of functions. For example, users would like to have additional digital inputs/outputs on decentralized inverters, as well as the option of diagnostics via fieldbus or a local communication interface.

As required for protection and switching technology, the inverters for controlling the motors are also expected to have a service life of eight to ten years, comparable to today's AC mains applications.

### Requirements for a Supply Technology

Supply technology plays a central role in the DC grid. It is responsible for the functionality of the grid and thus of production. Whereas up to now AC grid supply has largely been ensured by the energy suppliers and DC grids were only found in local contexts, this is changing regarding DC grids. The factory operator becomes the internal grid operator and therefore has to assume responsibility for the stability and robustness of the power supply. This important function prompts special requirements for the supply technology.

It should be possible to build up the feed from different modules. These modules in turn can be connected in parallel to supply the DC grid. Such a modular and partly redundant supply technology enables a safe operation and allows the growth of the supply along with the power demand of the factory. On the part of the supplying AC grid, the DC grid supply should filter the AC grid in a controllable manner and reduce harmonics. In addition, the DC supply shall take over the reactive power compensation on the AC side, thereby replacing an active line filter. In terms of power flow, the supply module is intended to be regenerative in order to feed self-generated energy into the AC grid, for example from photovoltaic systems. Self-generated energy cannot be used or stored otherwise in the DC grid. This feature also enables active influence on the public power grid by providing flexible energy from storage that can be injected into the grid when needed. Economic revenue concepts through flexible energy control of the energy supply are becoming conceivable. With respect to life expectancy of the DC grid supply, it competes with the longevity of a transformer in an AC supply. The later have a life expectancy of 20-30 years. Since such lifetimes are difficult to achieve with power electronics, the modular and partly redundant design of the devices is particularly important.

Even if the supply is referred to in terms of DC voltage, this should not be understood as a fixed, constant voltage. Rather, the voltage moves within a permissible voltage range (see Chapter 5.3). The lower limit of the operating range is directly dependent on the AC mains voltage. The other voltage limits must be adjusted accordingly. This is done in the system concept for the supply from 400 V AC mains as well as 480 V AC mains. In addition to providing a powerful DC voltage for drives and process technology, the supply technology also includes the 24 V/48 V extra-low voltage supply. In the future, this will also be supplied from the DC grid. Here it is important to consider the mains start-up separately. If the supply unit requires an extra-low voltage to build up the DC mains voltage, then this has first to be generated from the AC mains (in the supply unit itself or in the extra-low voltage power supply unit), as a supply from the DC mains is not yet possible.

In order to increase the efficiency of the supply units, it would be advantageous from the user's point of view if large supply units, consisting of several modules, had water cooling. The energy captured in the cooling water could be used, for example, to heat service water or for heating in general.

### Requirements for Cables and Plugs

Cables and plugs are used for power distribution in the plant and for connecting components to the DC grid. The main requirement with regard to cables is their electromagnetic compatibility (EMC). On the one hand, the cable routing should be robust against external effects of EMC radiation, and on the other hand, it should not emit any increased EMC radiation itself. Shielding of DC cables as a solution is out of the question, since this would lead to a significant increase in costs. Furthermore, the handling of shielded cables is more complicated during connection. It should be noted that in today's systems, both pre-assembled cables and on-site assembled cables are used. In today's AC-supplied systems, the communication lines as well as the motor lines are shielded.

With regard to the plugs, there should not be any danger to people or the system when disconnecting. This also applies to disconnecting under load. If a plug connection is not disconnected under load as intended, the plug connection may be destroyed; however, other system parts and people must be protected. To prevent such disconnections under load, plug connections can be interlocked. Due to the bidirectional energy flows, the plug connections must be designed in such a way that they are safe to be touched on either side. In addition, the plug connections must be provided with reverse polarity protection to prevent confusion between high-voltage (650 V) and low-voltage (5-60 V) DC systems and AC systems. More information on the protection of persons can be found in Chapter 7.3.

### Requirements for Management and for Energy Storage

Regarding the requirements for the management of a DC grid, a fundamental distinction ought to be made between energy management and grid management. Grid management is intended to ensure the load flow in the grid in such a way that power is balanced in real time between the power offered and the power required. In addition, grid management attempts to guarantee a safe and robust grid operation. Energy management essentially addresses cost-optimization and efficient energy provision. This means that nowadays the main responsibility for grid management lies with the energy supplier. Companies purchase their electricity from a supplier and primarily take on energy management tasks, supplemented by the required grid maintenance with regard to grid feedback effects such as harmonics or reactive power. In the future, a change will take place. The responsibility of grid management within the factory will shift to the operating company. Thus, it is important for the companies to be aware of the requirements for grid management and to evaluate and prioritize them for themselves.

In the study on which the project is based, securing the energy supply was named as an important requirement for grid management in combination with energy storage systems. For example, sufficient energy must be provided so that production plants can control a safe operating point in the event of a failure of the AC supply grid. A time of one minute is considered to be the lower limit. Control systems (e.g., PC or PLC) require a period of two to five minutes for data backup and controlled shutdown. Obviously, these must be protected for a longer period. Mains disturbances in the range of less than 200 ms should be bridgeable so that production may continue without any loss of quality. In the event of a power reduction required by the energy supplier, the grid management system must be able to draw the missing power from a storage facility or to throttle flexible production equipment or building supply systems such as ventilators.

If a fault occurs at a single grid participant, the entire station, i.e., the production-related grid of – in terms of process-logic – interconnected participants, can be switched off. The plant, which consists of several stations, then follows about 30-60 seconds later as soon as the processes have been terminated or safely switched off. Plants that are not directly connected to the plant in question should be able to continue to operate unaffectedly.

When defining the grid management, an essential requirement is that the configuration effort should be minimal. Planning and commissioning should be carried out with as little effort as possible and not more than the respective AC supply in production. This also applies to any grid reconfiguration tasks when adding or removing participants. In order to exclude sources of interference, the grid management should rely on as little communication as possible. For all necessary communication, existing fieldbus systems, such as PROFINET, EtherNet/IP or EtherCAT, should be used. In order to support the superimposed energy management, measured energy values, which are recorded within the scope of the grid management, are to be transferred back to the former. In the future, the users see a need for support regarding grid condition monitoring. They favor an integrated predictive maintenance approach that would ensure both a smooth and resource-efficient DC grid operation.

### Safety Requirements

It goes without saying that a future DC grid must meet at least the same safety standards as today's AC-based supplies. In addition to an earthing concept that includes all components and machine frames, this also includes a safety concept analogous to existing power supplies. Background information on how to design such an earthing concept in a DC grid can be found in Chapter 5.4. Such a safety concept and a detailed discussion of the challenges as well as an overview of the solutions in the various devices is the subject of Chapter 7.

**Other General Requirements**

A DC grid for the factory ought to be an open system. Components from different manufacturers have to interact smoothly with each other on the basis of common rules. At the same time, the DC grid should be scalable in terms of its functions. It must therefore be possible to operate a DC grid with simple, basic functionality. The necessary range of functions is to be determined by the application. This graduation of functions means that components with different functional scopes can also be offered. However, the compatibility of the basic functions has to be ensured. In this way, price-sensitive realizations of DC grids should be possible, too.

Implementing a DC grid means observing international regulations. These primarily come from the mechanical engineering environment. They ensure exportability of modern, DC-supplied equipment. In addition, the devices must be designed in accordance with the proven requirements regarding protection class and safety functions. In addition to these regulations, there are practical concerns, for example, the cable length in the plants should not constitute an obstacle to implementing a DC grid. It may consist of any number of individual lines, which are not limited in their total length. A length of 150 m was initially assumed for the total plant extension. Longer extensions of the grid are conceptually possible but may force additional measures concerning the reduction of oscillations or voltage drops.

Finally, one important requirement should be pointed out in particular. The DC grid is an open system that allows the user to integrate any generator, storage device or consumer. Conversely, this means that compatible devices must be developed and offered. At the core is the system concept with the basic functions of the grid management. Special functionality that offers optimization or other value-added services can be implemented on a manufacturer-specific basis.

## ■ 3.2 Conclusion from the Requirements

Overall, the requirements reflect a mix of technical needs, further development of technological trends toward decentralization, as well as modularization, usability and safety issues. If the requirements are evaluated separately by developers of electronic subsystems for a DC grid on one side and the users of such a grid on the other, significant differences emerge (Figure 3.2). The respective prioritizations are based on the subjective assessments expressed by the participants in the DC-INDUSTRIE project. Aspects of importance for the application, the difficulty and the challenge for the implementation are included in the prioritization. Although 69 of the 92 requirements are prioritized similarly, there are differences with regard to grid management and supply issues.

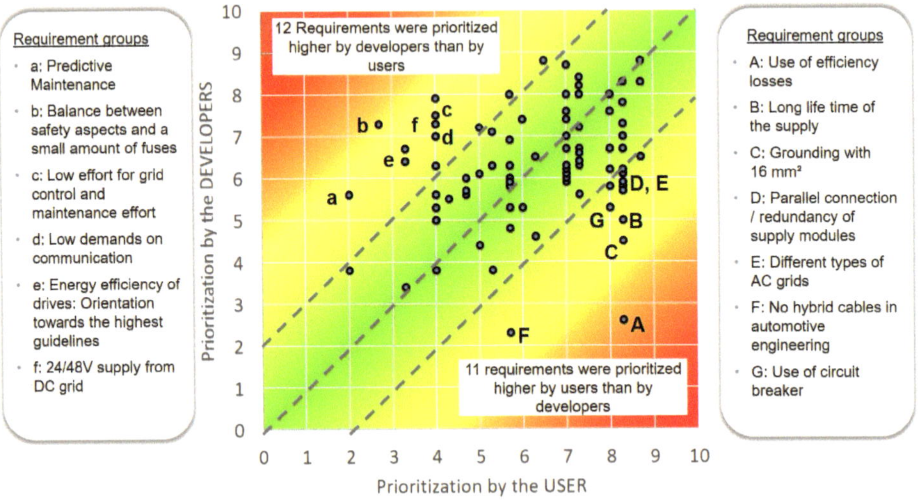


**Figure 3.2** Prioritization of requirements in developer and user perspective

The developers see focal points and challenges in the development of grid control. Particularly important are the requirements low communication effort, low-effort control and low maintenance effort. In this context, a low-effort control is comprehensible regarding its mode of operation. It is robust in terms of disturbances and changes in the grid. Maintenance effort describes the work required for operating the grid. This also includes the work that is to be carried out when the grid is changed due to additional or omitted participants.

In addition, the developers consider the balance between sufficient safety on the one hand and as few backup devices as possible on the other to be very important. The reason for this assessment is the expected technical effort for DC circuit breakers as well as their size and the resulting costs. The developers also attach greater importance than the users to the topic of predictive maintenance combined with underlying condition monitoring. The service life requirement for infrastructure components in the range of 10-20 years cannot be met with current power electronic devices. Since these devices are indispensable for the energy supply of production and thus for a smooth production process, modern concepts of maintenance are required, which must be adapted and further developed for power electronic devices.

From the user's point of view, other and different topics are weighted more. Energy efficiency and thus the reduction of efficiency losses is an important concern for users. This also includes water cooling with subsequent use of the waste heat for other processes. Existing technical solutions should be taken into account during implementation. A long service life is essential for the application. The reason for the different evaluation lies in the respective approach chosen. On the one hand, it is possible to design more conservatively, which would lead to an increase in equipment costs. On the other hand, predictive maintenance is seen as a solution module from the developer's point of view. This solution shifts the requirement for devices towards the system level. A parallel connection and thus a redundancy of supply modules is also seen in this context. Technically, this is hardly possible in a manufacturer-independent DC system. Developers see the challenge in a low-effort grid control, which is responsible for the parallel connection of the devices. Grounding, also in interaction with the different grid forms on the AC side, is a requirement, the implementation of which must be worked out within the system. Different grounding concepts have to be compared and evaluated with regard to the application. The aversion regarding hybrid lines, i.e., in which data lines and power lines are bundled, stems from the - supposedly - higher effort required for the connection. Here, attention must be paid to the ease of handling of the DC lines. They should by no means be inferior to AC lines. The requirement to dispense with fuses can be explained in a similar way. Maintenance personnel may have to check all fuses in the event of a fault. The correct fuses for replacement have to be at hand. This is time-consuming and logistically complex. Reclosers with circuit breakers or automatic circuit breakers are more convenient and can reduce downtime in the event of a fault.

Overall, it can be seen that a similar assessment of the requirements prevails. Differences can be identified above all regarding the approach and the handling effort. Based on these requirements, the next section will show where the strengths and weaknesses of DC technology lie in factory use. Also of interest are the obstacles that generally affect DC technology outside of technological development.

## ■ 3.3 SWOT Analysis of DC Technology

As part of an early Strengths, Weaknesses, Opportunities and Threats (SWOT) analysis, users were asked how they assess DC technology in their area of application (Figure 3.3). The focus was not on the technical details of individual devices, but on the system concept. The reference here was also the existing AC supply grid in the factory.

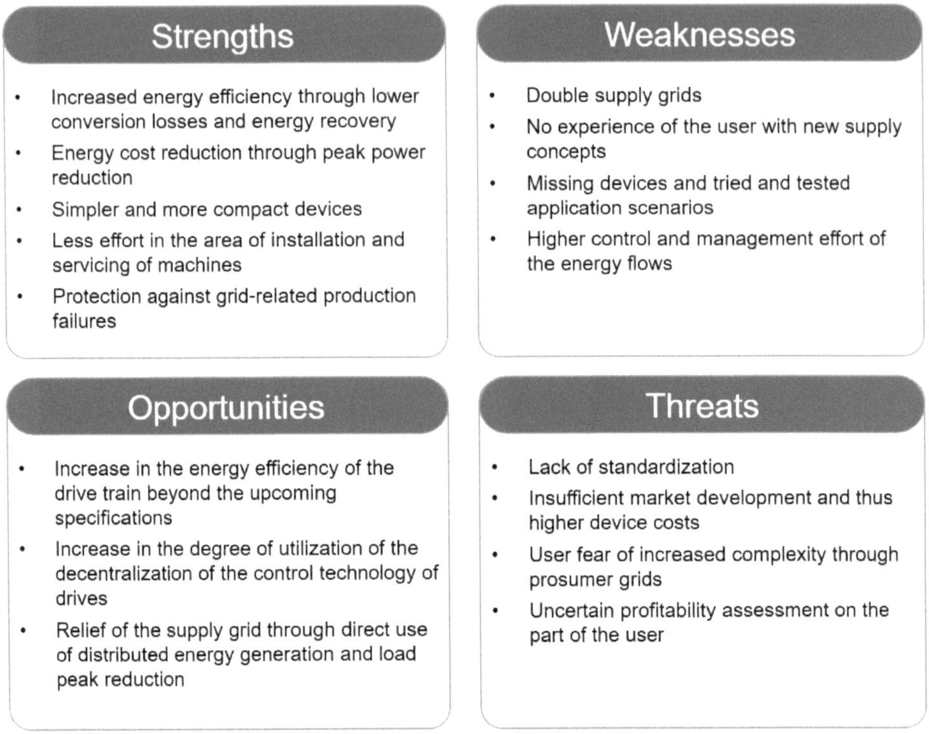


**Figure 3.3** SWOT analysis of DC supply for production facilities

The **strengths** of the DC grid lie in the areas of energy, material and production efficiency. Cost savings and energy demand reductions are associated with the use of a vendor-independent DC supply in the factory. Both are achieved through an increase in efficiency, coupled with a reduction in peak loads, which is made possible by the direct exchange of energy between the machines and the integration of storage systems. Further cost-saving potential is linked to simplified installation of the machines and a robust and fail-safe production operation. The latter is also related to the storage capability in the system as well as the use of energy from purposefully decelerating machine parts in case of supply disturbances. The robustness is reinforced by system-inherent filtering of harmonics or flicker from the AC grid.

The strengths are offset by the **weaknesses**. Particularly in a transition period, which can sometimes last ten or more years in established factory structures, two parallel supply grids – AC and DC – are necessary. This would mean additional investment costs and increased maintenance. As with all new technologies, there is a lack of experience in the field of DC supply for factories. This starts with the planning and design of the DC grid and extends to the behavior of maintenance personnel when dealing with new DC equipment. In addition to little application experience, there is also a lack of equipment to install a DC grid that meets demand. Available devices do not currently come from the production environment, but rather originate from application areas in which DC technology has already been established for years, such as photovoltaic systems, railroad and ship technology, or data centers. Finally, the necessary effort for grid management that will have to be put in by the factory operator in the future is unclear - and not only is it unclear, it is also feared to be high. Today, the factory operator's tasks with regard to grid management consist of filtering out disturbing influences such as harmonics and reactive power. Due to the unidirectional flow of energy from the power plant via the supply grid and transformer into the factory grid to the machine, there is no need for coordination of power flows. It is only the use of storage facilities and power self-generation plants that increases the effort. The possibility of a permanent energy exchange with a bidirectional power flow, which is also variable in direction over time, increases the complexity and thus also the supposed effort for grid management.

Notwithstanding the weaknesses that a DC grid can bring, especially in the early years of technological change, this technology shift does offer **opportunities**. One is the many opportunities it offers for drive technology, as discussed in detail in Chapter 4. The isolated view of the efficiency of an electric motor and the effort to optimize it in isolation, threaten to lead to a dead end. The system perspective opens up additional efficiency potentials associated with a DC grid. Here, there is an opportunity to take a decisive step forward. In addition, the reduction in power electronics and the lower cooling requirements due to the elimination of rectifiers and some of the DC link capacitors in the frequency converters offer the possibility of relocating these electronics directly in the motor. This allows more modular machine designs. Another opportunity presented by the introduction of DC grids in industry is to relieve the strain on public supply grids and thus support the energy transition. If it is possible to efficiently use or store generated energy where it is produced, then the transmission grids will be relieved. In addition, there is the possibility of making energy use or energy feed-in more flexible. The public supply grid is not only relieved, but actively supported. The decentralized control of the DC grid distributes the energy in the grid and integrates storage efficiently and without expensive infrastructure. Thus, the decentralized control offers the potential of a direct coupling with the energy management system. In this way areas down to the machine level could be controlled in a low-effort and decentralized energy-flexible manner.

Of course, the **risks** associated with an early switch to DC technology should not be concealed. A lack of standardization always carries the risk that system components will not be 100% compatible and interchangeable at a later date. It should be noted at this point that many standards already cover both, AC as well as DC systems. The expertise in the DC-INDUSTRIE research project and subsequently in

DC-INDUSTRIE2 has been chosen broadly and across manufacturers so that the solutions found are technologically sensible and future-proof. Building on these results, standardization activities are currently being initiated at national and international level. A further risk of early use is the underdeveloped equipment market and the uncertainty about how it might develop. One consequence could be permanently expensive components. On the other hand, there is the question of which technological alternatives to DC technology are available for an industrial energy grid. Significant changes in energy supply will be needed to tap into existing potentials and open others. For example, there will be entirely new factories with DC supply, but also established factories in which DC supply is only gradually being introduced. This results in two parallel approaches for the development of the market, which can in turn favor the demand for devices from different suppliers.

The increase in complexity in the planning and design as well as in the subsequent operation of a DC grid cannot be dismissed at first. In the future, the factory operator will take over tasks that the energy supplier does for him today. Irrespective of DC technology, energy supply in companies is becoming more complex. Variable electricity tariffs require more flexibility in order to obtain energy economically. This requires the use of storage and self-generation of energy. The difference to the complexity of an AC grid can be minimized via planning and design methods. Finally, there is uncertainty about the economics. The estimates in Chapter 2.5 give a first overview. It can be seen that even at the beginning of the technology change, the cost difference in the devices is small and an economy of scale occurs early. This does not take into account the potential on the profit side. If the complexity in the energy grid increases due to storage and generators, the DC grid has a clear advantage.

In summary, the strengths and weaknesses as well as opportunities and risks must be evaluated in an individual perspective taking into consideration the production facilities, location and public supply grid of each individual company. The interest of a wide variety of user companies already shows that the idea of rethinking the energy system of the factory is important and purposeful. The DC grid offers attractive solutions here.

# 4 Opportunities for Drive Technology

Karl-Peter Simon
Simon Puls
Jan Schupp
Karlheinz Wirsching

The continuous focus on higher energy efficiency and sustainability is and remains an important driver for the implementation of the energy transition. In industry, 70% of electricity consumption is accounted for by electric drives [Kosc2017]. This makes drives by far the largest consumer of electrical energy. Any reduction in the power consumption of drives through efficiency improvements also means a corresponding reduction in $CO_2$ emissions. Since January 1, 2017, all new three-phase motors sold in Europe in the power range from 0.75 to 375 kW must meet the requirements of the energy efficiency class IE3, or IE2 if operating frequency converters. From 2021 onwards, this exemption will cease to exist, thus only IE3 motors may be used [Comm2019].

However, the previous standards and norms for increasing energy efficiency have somewhat failed to initiate a rethinking and to introduce truly ecological solutions. The legal regulations mentioned have so far only been applied in the context of operating a power grid for three-phase motors. The definitions of the legal efficiencies have been stipulated at rated speed and rated torque, in other words, at an operating point that is of little practical significance. For such motors, the cost increases due to a much higher use of material, for example copper. The effect of saving energy is not present in many practical applications. The approach of examining the entire drive system makes it possible to reverse this trend and to move towards energy-efficient and economic solutions. A DC grid provides the basis for such solutions. In this interconnected grid, we have the freedom to optimize electric motors for an operation with frequency converters and to enable energy exchange within the drive systems. How this works will be explained in more detail in this chapter.

The introduction of DC technology is intended to - for the first time - optimize system costs while at the same time to further increase energy efficiency. The grid structure is based on an AC voltage feed-in, which provides the DC voltage supply for the production plants via a central rectifier. Active line filters are integrated in the central rectifier to ensure the necessary voltage quality. Frequency converters usually have an integrated rectifier with mains filter, which always means a lossy conversion of the electrical energy from AC to DC voltage. By supplying the frequency converter directly with DC voltage, all decentralized energy conversions can be eliminated. Since centralized energy conversion (from AC to DC) is much more efficient, conversion losses are significantly reduced. By supplying all electric motors directly with DC voltage via a frequency converter, all installed motors are connected via a common DC voltage grid. This allows direct energy balancing of all accelerating and braking drives. The central DC voltage supply only feeds-in the differential energy. It only has to convert this from AC voltage to DC voltage. Additional components, such as braking resistors, which burn up energy, are eliminated. This means that fewer components have to be installed and space is saved. A DC voltage grid essentially causes only ohmic transmission losses. Compared with an AC grid, there are no capacitive and inductive line losses. This means that the conditions are in place to significantly increase the use of frequency converters. Electric motors will increasingly no longer be optimized for a 50 Hz grid, but for optimum use in operation with frequency converters. This will enable much better utilization of the asynchronous machine and thus in turn will reduce costs and increase energy efficiency.

Since a frequency converter fed with DC voltage contains fewer components (omission of AC line filter, rectifier, braking resistors and part of the DC link capacitors) its size is smaller thus allowing an easier integration into the motor. In addition, this eliminates the cost of expensive shielded motor cables. Especially the high capacitive losses of the motor cables [Wint2017] between converter and motor are no longer relevant in such structures – in contrast to today's system designs with long motor cables. Today, about 35% of all newly sold asynchronous motors are operated with an additional frequency converter in order to be able to continuously adjust the motor speed to the current demand. A DC grid makes it possible to significantly increase the proportion of variable-speed drives, which is urgently needed both for implementing Industry 4.0 applications and for effectively increasing energy efficiency.

## ■ 4.1 Challenges and Operating Principle of an AC Grid

In machines, such as those found in the packaging or plastics industries, the movements of the machine parts are predominantly discontinuous. This means that the machine parts are not moved uniformly or constantly at one speed, but repeatedly perform start and stop movements. As a result, the motors used have to be accelerated and decelerated very often. If the motors were connected directly to the industrial grid, depending on the design of the motor, for example, only a constant rotary motion of 3000 rpm would ever be possible. One reason for this is that the existing industrial grid has a constant frequency of 50 Hz. In order to be able to manage the speed changes, the frequency of the motor must be changed. For this purpose, a frequency converter is required, which consists of the supply, DC link, converter and the smart controller (Figure 4.1).

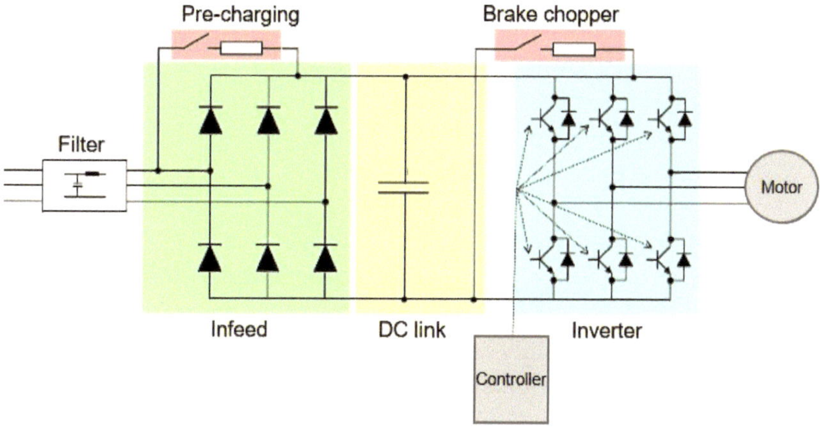


**Figure 4.1** Topology overview of a frequency inverter

The feeder is connected directly to the 50 Hz main. It rectifies the AC mains voltage. This creates a DC voltage in the so-called DC link. Incidentally, the feeder usually consists of three diode bridges, each of which rectifies one phase of the 3-phase grid.

The converter consists of isolated gate bipolar transistor modules, or IGBT modules for short. Their task is to modulate the DC voltage in such a way that an AC voltage with a certain predetermined frequency is produced. The level of the frequency is determined by the smart controller. By controlling the IGBT modules, this controller can generate a variable frequency at the output of the inverter, i.e., at the connection of the motor. This allows to change the speed of the motor in relation to the frequency converter, and – in other words – dynamic movements of the machine parts.

However, a frequency converter alone is not enough. The system requires further accessories: On the one hand, these are mains filters. In order for the frequency converter to generate stable output frequencies, the mains voltage should not be contaminated, i.e., interference on the mains caused – for example – by other electronic devices should reach the input of the frequency converter or the feed as little as possible. Therefore, a mains filter is installed upstream of each device so that it filters out the mains interference. At the same time, it has the task of filtering out interference generated by the frequency converter itself so that these do not contaminate the mains - examples would be switching peaks of the IGBT modules.

Secondly, capacitors must be installed in the link system of the converter. They serve to attenuate ripples in the DC voltage caused by the feed-in and also to mitigate interference caused by the converter.

Basically, a motor that is decelerated works like a generator. It generates energy that is fed back into the DC link via the converter. The DC link has only a limited capacity to absorb energy due to the built-in capacitors. In order not to exceed this storage capacity and possibly destroying the device, braking resistors are connected to the DC link. The braking resistor is switched on depending on the level of energy in the DC link. It converts the generated energy into heat. From an energy point of view, this is not a sensible solution, as energy is uselessly converted into heat. To avoid this, there are different variants of the standard frequency converter on the market, as described above. For example, instead of a pure feed-in, a so-called feed-in/feed-back unit (= Active Infeed Converter) can be installed in the frequency converter. This feeds surplus energy that has been stored in the DC link back into the grid. Another possibility is that not only one inverter is connected to the DC link, but several. In this way, energy that enters the DC link through decelerating the motor can be "sucked out" of the DC link again with the help of another converter. The advantage: no more energy has to be converted into waste heat.

Frequency converters have an energy efficiency of 90-98%. This depends on the power of the respective frequency converter; smaller converters have a worse energy efficiency than larger ones.

Looking at the overall system (line filter, frequency converter, braking resistor, etc.), the energy efficiency can be increased even further. Energy can be saved especially by smart architectures of the overall system. The two biggest influencing factors here are the DC link grid with several converters connected and the central regenerative power supply, which no longer requires a braking resistor. However, today such an interconnection is not possible across all manufacturers.

## ■ 4.2 From AC to DC Grids

The following section deals with how the changeover from a 3x400 V AC system to a 650 V DC system affects efficiency and material usage. A DC system can deliver up to 20% more power if the material input is left the same for both variants. Conversely, it is also possible to create a system with the same power output and save up to 20% in material.

Furthermore, in an AC system the energy flow can only be in one direction. In situations where mechanical loads are decelerated, a conversion of mechanical energy into electrical energy takes place within the motor. However, this cannot be consumed in a useful way. On the contrary, additional components are required to convert this excess energy into heat. In a DC system, on the other hand, energy can be exchanged between devices and used by each one.

### Objective of the DC Grid Approach for Drive Technology

The purpose of electric drives is the conversion of electrical energy into mechanical energy. This energy conversion should be as efficient as possible and at lowest possible investment costs.

When converting from AC to DC technology, these goals can be achieved by the following measures:

1. Conversion from the decentralized rectification and filtering in the individual drives to a centralized generation and filtering of the DC voltage
2. Coupling of the link circuits and a possibility of energy exchange between clocked devices
3. Making use of the high voltage level in the DC technology
4. Consideration of the simultaneity factor of the loads
5. Conversion of all AC drives running synchronously with the mains to converter control

### State of the AC Technology and Envisaged DC Technology

Figure 4.2 shows the state of the art of AC technology: Individual drives with and without converters are operated on the AC grid. There is no energy exchange between the individual drives, excess braking energy is burned up in braking resistors. A large number of frequency converters generate a considerable amount of feedback into the AC mains. Each converter generates its own DC voltage internally. The mains feedbacks generated at the local AC connection must be compensated by filters.

No energy exchange takes place between the individual processes. Excess braking energy is converted into heat in braking resistors. This regenerative energy share is typically between 0% and 5%, on average 3% of the total energy of the drive. Mains synchronous drives are not controllable, but they feed energy back into the AC mains during braking while in regenerative operation mode of the motor. In rare cases, regenerative converters are used.

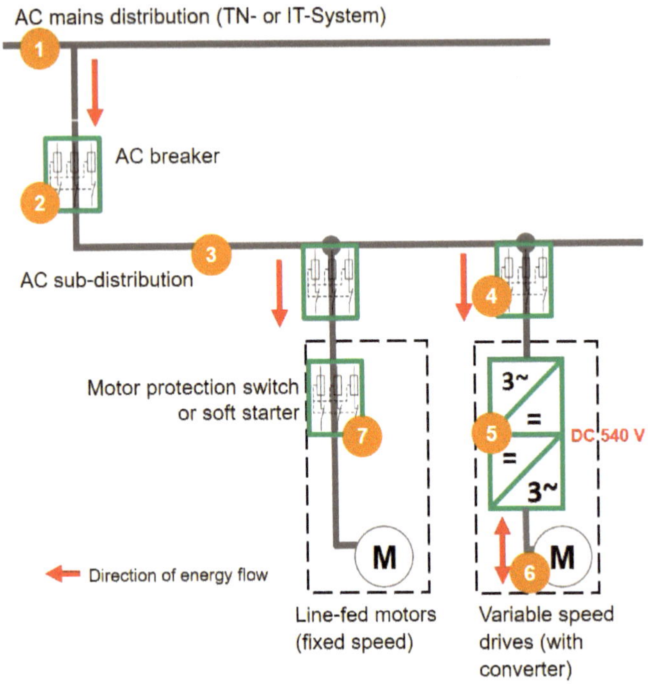


**Figure 4.2** State of the art AC technology (single drive on 3x400 V AC grid)

Advantages of AC technology:

- Structure of the grid with regard to topology, fusing and wiring is known
- Line-commutated drives can be easily integrated (currently approx. 65% of all motors)
- Mains-operated drives are unproblematic with regard to their braking energy (sinusoidal voltage)

Disadvantages of AC technology:

- With converter-operated drives, braking energy is converted into useless heat
- Depending on the size of the plant and the proportion of speed-controlled drives, complex compensation systems have to be designed at the connection point of the main AC distribution and the grid operator

Figure 4.3 shows the envisaged technology for a DC grid in the factory:

Drive groups and individual drives are operated on the DC grid. Energy is exchanged between the individual processes, and excess braking energy is used in other processes. Grid feedback on the AC side is reduced by central AC-DC converters with integrated filters.

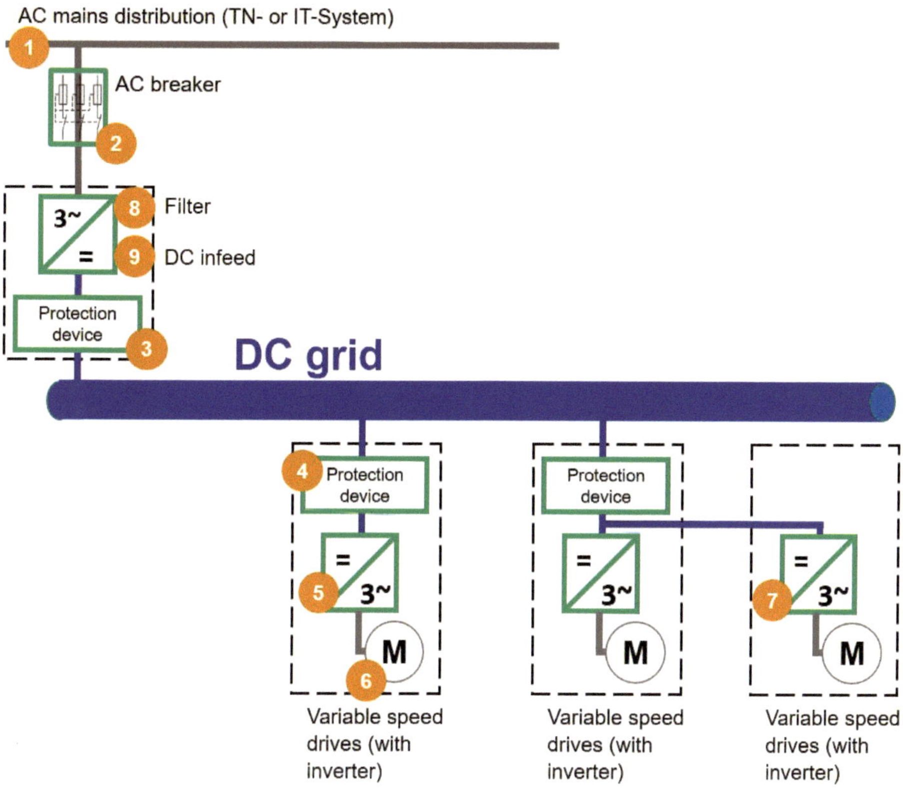


**Figure 4.3** Conversion of the system to a 650 V DC grid

Advantages

- Central AC-DC supply and filtering is effective and inexpensive
- Energy exchange allows braking energy to be used in other processes
- All drives are controllable and thus adaptable to the current load

#### Technical Evaluation of the System Blocks

In this section, the blocks from Figure 4.2 are compared individually with the blocks from Figure 4.3 and evaluated. The numbering corresponds to the numbering in the figures.

#### 1. Main Distribution

DC-INDUSTRIE, at the scaling level shown here, requires an IT system to be assigned exclusively to the supply of the DC grid. For large systems, the grid connection transformer that is present anyway takes over the function, so the effort is the same. For other configurations, an additional transformer may be required.

### 2. AC Main Switch and Fuse

The effort is the same.

### 3. Wiring in the Sub-Distribution

Compared to AC technology, DC-INDUSTRIE requires fewer cores in a cable for energy transmission. This means that the conductor cross-section of each core can be increased while using the same amount of copper. Furthermore, the voltage level is higher with DC. In addition, there is an increase in the transmittable power due to the elimination of reactive power with DC. These effects mean that more power can be transmitted with the same amount of copper or less copper is required for the same power.

**Example: In a 4-wire system in AC**, three conductors and the protective earth (PE) are required. In DC, two conductors plus PE are always required. Due to the small number of conductors, the conductor cross-section $A_{DC}$ can be increased by 33.33% in DC without additional copper consumption (Equation 4.1). This also leads to a 33.33% higher transmittable current in DC.

$$4 \cdot A_{AC} = 3 \cdot A_{DC} \tag{4.1}$$

$$A_{DC}/A_{AC} = 4/3 = 1{,}333$$

If the AC system additionally contains a neutral conductor (5-conductor system), then the cross-section for DC can even be increased by 66.67% in comparison. This consideration applies to conductor cross-sections up to 16 mm², where the cross-section of the PE conductor is the same as that of the outer conductors. For conductor cross-sections >35 mm², the required cross-section of the PE conductor is 50% of the outer conductor cross-section [IEC60364]; the calculation for this then provides a 40% higher current in the 4-wire AC system due to the 40% higher conductor cross-section $A_{DC}$ compared to $A_{AC}$ with constant copper input. Furthermore, the phase voltage in AC is 230 V, while the comparable virtual phase voltage in DC is 325 V (half the nominal DC voltage), so it is 41% higher in DC. However, it must be noted that three strings are available for power transmission in AC and only two strings in DC.

The transmittable power of a line is calculated as follows:

$$P_{Leitung} = U_{Strang} \cdot I_{Strang} \cdot N_{Strang} \cdot \cos \varphi \tag{4.2}$$

Here, the RMS values of current and voltage are to be used for AC voltage. The ratio of the transmittable power between DC and AC results for power factor $\cos \varphi = 0.8$ (typical value for the consideration of the reactive power of asynchronous machines or converters) and for cross sections up to 16 mm² according to Equation 4.3 to:

$$P_{\frac{DC}{AC}\%} = \frac{U_{Strang\ DC} \cdot I_{Strang\ DC} \cdot N_{Strang\ DC}}{U_{Strang\ AC} \cdot I_{Strang\ AC} \cdot N_{Strang\ AC} \cdot \cos\varphi} \times 100\ \% = \tag{4.3}$$

$$= \frac{325V \cdot 1{,}3333 \cdot 2}{230V \cdot 1 \cdot 3 \cdot 0{,}8} \times 100\ \% = 157\ \%$$

Thus, in this case ($\cos\varphi = 0.8$) 57% more power can be transmitted with the same material input (Figure 4.4 ), and even 64.8% for cross-sections above 35 mm$^2$ with the factor 1.4 (instead of 1.33) in the above equation. Alternatively, it is just as possible to save 36.3% copper for cross sections up to 16 mm$^2$ with the same transmitted power (due to copper saving equal to $1 - P_{AC}/P_{DC} = 1 - 100/157\ \%$). Furthermore, the lower number of conductors also reduces the amount of assembly work required when building the system.

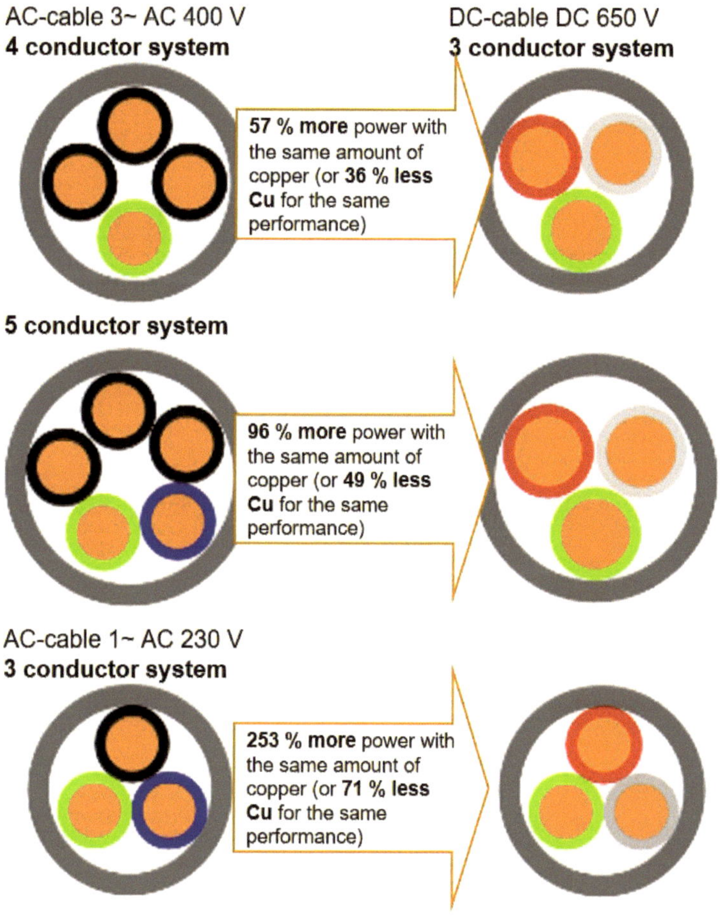


**Figure 4.4** Comparison of the transmittable power between AC and DC at the same copper cost (for cross-sections up to 16 mm$^2$ with a power factor $\cos\varphi = 0.8$)

### 4. Protection and Switching Device in the Sub-Distribution

In the DC system, special DC switches are required. Compared to AC technology, the additional cost is considerable. On the other hand, it is important to note that these switchgears offer a significant increase in functionality, such as controllability via programmable logic controller (PLC), reclosability after faults, precharging of capacitances, and measurement of current, voltage, and energy. If such smart grid functions were to be incorporated into AC, the effort would be comparable.

### 5. Drive Frequency Converters and DC Inverters

While AC uses frequency converters that are not capable of regenerative braking, DC inverters are capable of regenerative braking on principle. This means that braking energy can be used.

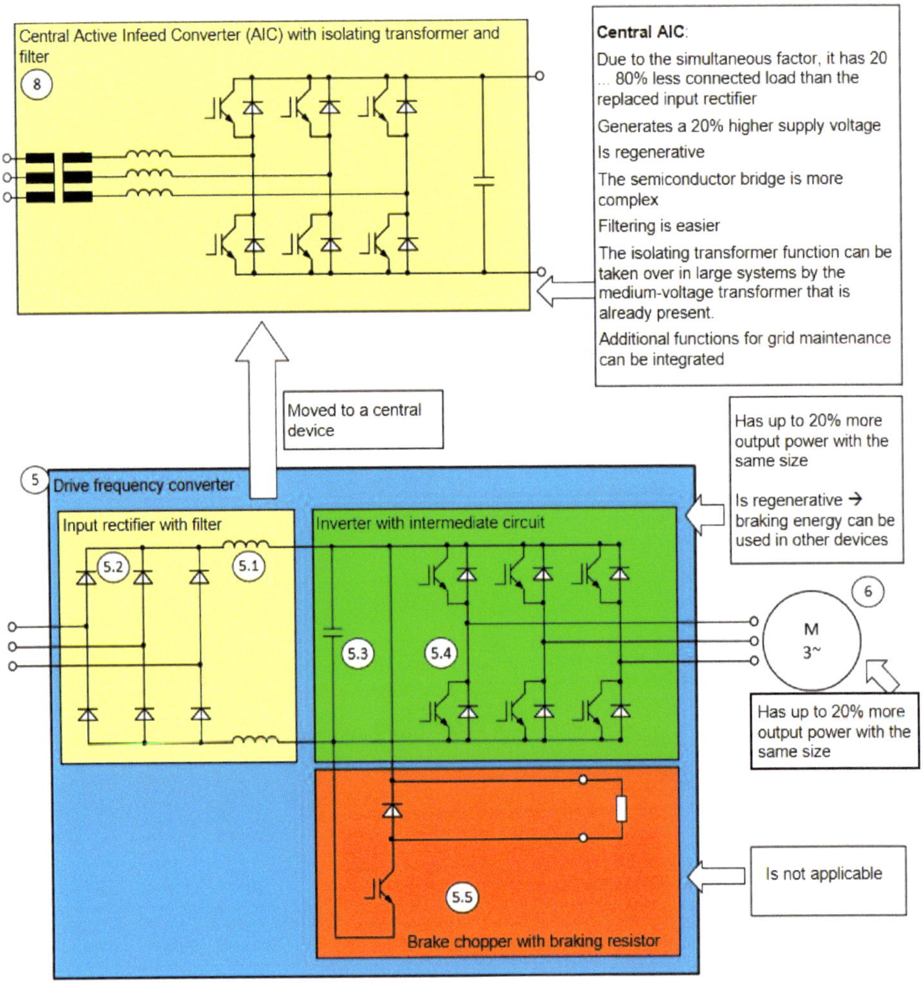


**Figure 4.5** Frequency inverters during the changeover to DC technology

Another advantage, however, is that by increasing the DC link voltage from DC 540 V to DC 650 V, the output voltage of the inverter is also increased by 20%, while the permanent output current is not reduced quite as much with the same components. Thus, one achieves a higher output power of the inverter, plus 5 to 10% are realistic values.

Figure 4.5 gives an overview of the conversion of an inverter to DC technology.

**5.1. Mains filter**

The filter against harmonic distortion is omitted in the inverter and is instead implemented centrally in the DC supply. The filter against common-mode interference is retained in the inverter.

**5.2. Rectifier**

Not used in the inverter but implemented centrally in the DC supply.

**5.3. DC link**

Due to DC link coupling, the harmonics of the inverters, which load the DC link capacitors, are added geometrically, while the capacitance values of the individual inverters are added arithmetically. This has the effect of reducing the load on the DC link capacitors by around 30%. Accordingly, the capacitors could therefore be designed smaller. On the other hand, the design must now take the weakest part into account, which means that some manufacturers cannot utilize their capacitors as usual but must consider the performance of other manufacturers. DC-INDUSTRIE has developed a specification for this.

**5.4. Inverter**

No difference between AC and DC.

**5.5. Brake chopper and braking resistor**

This assembly is not required for DC; the excess braking energy is used by other consumers in the DC grid. This means: less effort and at the same time increased efficiency.

### 6. Motor

With respect to the motor, the biggest difference is that the inverter can provide up to 20% more voltage with DC supply. This can be used by a same sized motor to generate up to 20% more power.

Example: A motor manufacturer specifies the following rated operating point for a permanent magnet synchronous motor in size 72M for a supply voltage of AC 400 V:

$N_{nom}$ = 2000 rpm; $M_{nom}$ = 30.7 Nm; $P_{nom}$ = 6.43 kW, the associated current is $I_{nom}$ = 13.2 A.

For AC 480 V, the following operating point is given for the same motor:

$N_{nom}$ = 2500 rpm; $M_{nom}$ = 28.7 Nm; $P_{nom}$ = 7.52 kW, the associated current is $I_{nom}$ = 12.3 A.

In the example, this increases the power of a motor by 17%.

#### 7. Asynchronous Motors Operated Synchronously with the Mains

This type of motor will no longer exist in a DC grid. Therefore, the goal would have to be to provide simple inverters for the respective applications that cost no more than a motor circuit breaker or soft starter.

On the other hand, replacing grid-synchronous motors has a significant energy saving potential, since the motor power can be controlled and thus adapted to the real demand of the application. One can even increase the motor power by increasing the frequency.

#### 8. Central Mains Supply

The central DC supply now takes over the function of blocks 5.1 and 5.2. but also includes an intermediate circuit. The great advantage of a central supply is that the simultaneity factors of all loads can be taken into account. Usually, depending on the application, simultaneity factors of 0.2...0.8 can be calculated. This means that the DC feeder can be 20...80% smaller than the sum of the rectifiers of the individual drives in AC technology.

Another advantage is the filtering, because the central filter has to be tailored to only one device. There is no additional interaction of the filter with a multitude of other devices and filters.

In the scaling shown in the figure, a high-quality Active Infeed Converter (AIC) is used. Compared to the passive rectifiers of the single drives in AC, the cost is higher, but this significantly reduces the overhead in the filter. The AIC raises the DC voltage to the 650 V as discussed above, giving the user higher system performance.

### Comparison at System Level

As shown in the previous considerations, the DC approach is advantageous from an energy point of view because it generates fewer losses and braking energy can be used in a meaningful way.

Regarding investments for system construction, it can be noted that in the area of energy conversion, i.e., DC feed-in, inverter and motor, the DC approach is advantageous because components are saved, the expenditure for filters is reduced while at the same time a higher output power can be delivered. The capital investment for distribution is also lower because, as has been shown, more power can be transmitted with the same amount of copper. However, the effort for contacting and switching is higher in the DC grid. It should be noted here that, compared to conventional AC technology, the plant operator gets a considerable increase in functionality, such as fieldbus connection of the protection devices and thus automatic switching, measuring and monitoring of the plant status. The reduction in losses also lowers the cost of air conditioning the control cabinets and systems.

Overall, the DC approach to drive technology makes a lot of sense.

## 4.3 Technical Evaluation

### Evaluation of Individual Devices

In the course of the DC-INDUSTRIE project, the participating drive manufacturers provided electric drives for the model plants at Daimler, HOMAG and KHS. The drives, marked as development samples, have been equipped in such a way that - built-in the DC terminals - they behave in conformity with the DC-INDUSTRIE specification (hereafter also referred to as “DC-INDUSTRIE-compliant”). In addition to the DC voltage, this means that, in contrast to standard AC and DC devices, no precharging of the DC link storage is necessary, since the precharging is solved centrally. Furthermore, components for controlling and connecting braking resistors are omitted, since regenerative energy is basically to be “fed back” into the DC grid. In order to limit the emission of ripple currents into the DC grid, components such as filters and DC link capacitors had also to be designed accordingly.

Only in such a way is an open, manufacturer-independent DC grid possible.

Electric drives consisting of an inverter and an electric machine can be both control cabinet and field devices. If several drives are connected together to the DC grid via a connection, this common connection must be DC-INDUSTRIE-compliant in all respects.

Figure 4.6 shows an example of the relative space required for the components in an AC device and the extent to which a DC device can be smaller, or by how much the power can be increased while maintaining the installation space. These include, in particular, varistors for protection against transient overvoltages from the supply grid, for example due to lightning, and a filter for compliance with EMC limits, consisting of a common-mode choke in conjunction with X and Y capacitors of an increased protection class. Rectifiers and precharging circuits, as well as brake chopper circuits, are also eliminated in drives for DC-INDUSTRIE. Corresponding to the reduction of components, the necessary installation space is reduced, and the losses become smaller.

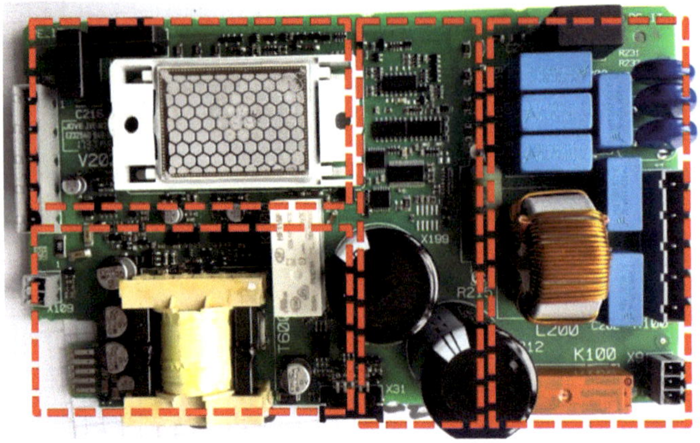

**Figure 4.6** Standard AC drive consisting of AC input circuit (right), DC link (center), power supply (bottom left) and inverter (top left)

What is not taken into account is, that a DC device must also have filters for compliance with conducted interference emissions and that the DC link capacitors become smaller in terms of their capacitance and in terms of the installation space required.

For an open DC grid, i.e., a grid with the possibility of using devices from different manufacturers, it is necessary that the connected devices have a minimum degree of robustness against loads, e.g., ripple currents. On the other hand, however, and indeed necessary, is that the connected devices do not overload the grid and in consequence the other devices that are operated in the grid. Due to the low-impedance DC coupling of the intermediate circuits of the devices via the DC grid, there is the technical possibility that ripple voltages in one specific device lead to equalizing currents, which are fed by the intermediate circuit storage of one or more other devices. To exclude the possibility of one device overloading another, limits have been set for the interference emission of ripple voltages. For this purpose, a test circuit (Figure 4.7) has been designed which can be used to test both a device in the DC grid and a single device on a DC grid replica.

The current going into the test circuit can be easily measured with a multimeter that has a sufficiently high bandwidth. The test circuit is connected directly to $+U_{DC}$ and $-U_{DC}$. By connecting the capacitor and resistor in series, a frequency-dependent impedance is formed which corresponds to that of an electrolytic capacitor. A single device operated on a mains replica may generate a maximum RMS current of 40 mA in the test circuit. If a grid of several devices is to be tested, a maximum current of 28 mA may be generated in the test circuit. The limits can be complied with in particular by using a sufficiently large DC link capacitance. As a guide value, 25 μF per kW DC power can be assumed here.

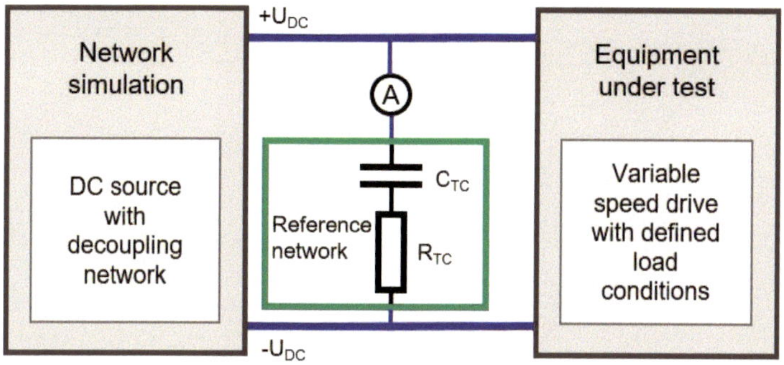


**Figure 4.7** Test circuit for checking a single device incl. its electrical arrangement

### Evaluation of a Device Grid

One of the most important advantages, due to the elimination of rectifiers, is the possibility to automatically feed regenerative energy back into the DC grid - without losses in components. Figure 4.8 shows the voltage and current curves as well as the resulting power curve. The negative power components illustrate the regenerative potential of the DC grid. Especially in the case of intermittent drives, for example robots, usable energy is available frequently and regularly. In contrast to the AC grid, in which this energy must be dissipated by braking resistors to protect the devices from overvoltage, the energy is thus automatically made available to other devices all while the energy requirement from the supply grid is reduced. The efficiency of the entire plant is increased.

The lower cost of designing and using variable-speed drives, even for decentralized applications, greatly expands the possible uses of electronic drives. AC mains motors are losing their raison d'être from a technical as well as an economic and ecological point of view. The use of inverters on an increasing number of drives simultaneously distributes a large number of sensors that can be accessed via fieldbus systems. Data, which each inverter collects anyway through the sensors necessary for its function, can be made generally available, for example for grid management or for simply monitoring a complete drive train including the driven mechanics.

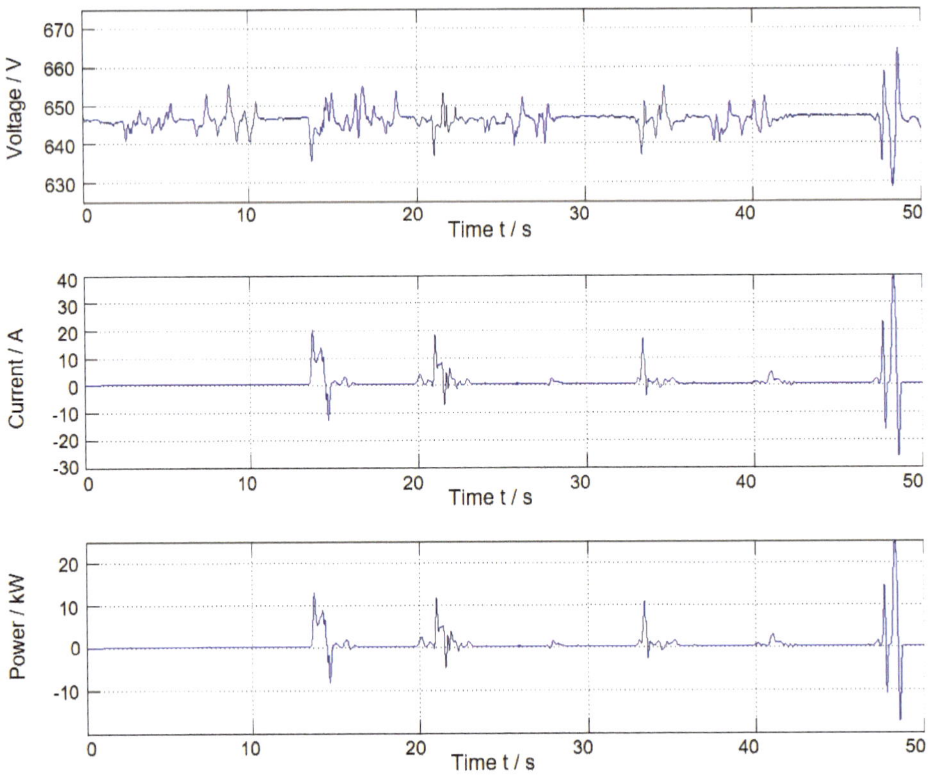


**Figure 4.8** Voltage (top), current (center), and calculated power (bottom) of a robot over one motion cycle

## Literature


[Kosc2017] Koschnick, Gunther; Sattler, Bernhard: Energieeffizienz mit elektrischen Antrieben. ZVEI – Zentralverband Elektrotechnik- und Elektronikindustrie e.V. Fachverband Automation Fachbereich Elektrische Antriebe, Frankfurt am Main, 2017

[Comm2019] Commission Regulation (EU) 2019/1781: Laying down ecodesign requirements for electric motors and variable speed drives pursuant to Directive 2009/125/EC of the European Parliament, https://eur-lex.europa.eu/legal-content/EN/TXT/PDF/?uri=CELEX:32019R1781&from=EN (access on 25 Apr 2020)

[Wint2017] Wintrich, Arendt: IGBT Schalten mit Kabellast. SEMIKRON Elektronik GmbH & Co. KG., Nürnberg, 2017

# 5 System Concept of a DC Power Supply

Martin Ehlich
Johann Austermann
Frederic Blank
Holger Borcherding
Benno Weis

In AC technology, any topologies are possible, which the user designs, installs and commissions. There are basic rules for the design of safety concepts (prospective short-circuit current, residual-current circuit-breaker, motor circuit-breaker etc.) and disconnection devices (e.g., emergency stop circuit), which are observed by every electrician. Furthermore, standardized high-voltage tests are carried out as part of the release process of machines and systems.

The aim of the system concept is to define the simplest possible rules for setting up safe and manageable DC grids so that the inhibition threshold for introducing this technology among component and machine manufacturers as well as plant operators is low. As far as possible, design rules, safety mechanisms, circuits and components known from the AC world are to be used.

In terms of application, the DC grid is intended for factory automation. Scalability from individual machines to complete systems to an entire factory floor is a key requirement. All electrical consumers, storage units and generators are energetically coupled via the DC grid. In a factory, this includes the production and process technology systems as well as the infrastructure of the factory buildings, i.e., lighting, heating, air conditioning and ventilation.

As part of the "DC-INDUSTRIE" project funded by the German Federal Ministry for Economic Affairs and Energy (BMWi), a system concept has been developed that enables the operation of different devices across various manufacturers if they adhere to the specifications agreed upon in the project. Compliance with this specification is referred to in the following as "DC-INDUSTRIE-compliant".

## 5.1 Topology in the DC Grid

Figure 5.1 shows the basic structure of the industrial DC grid.

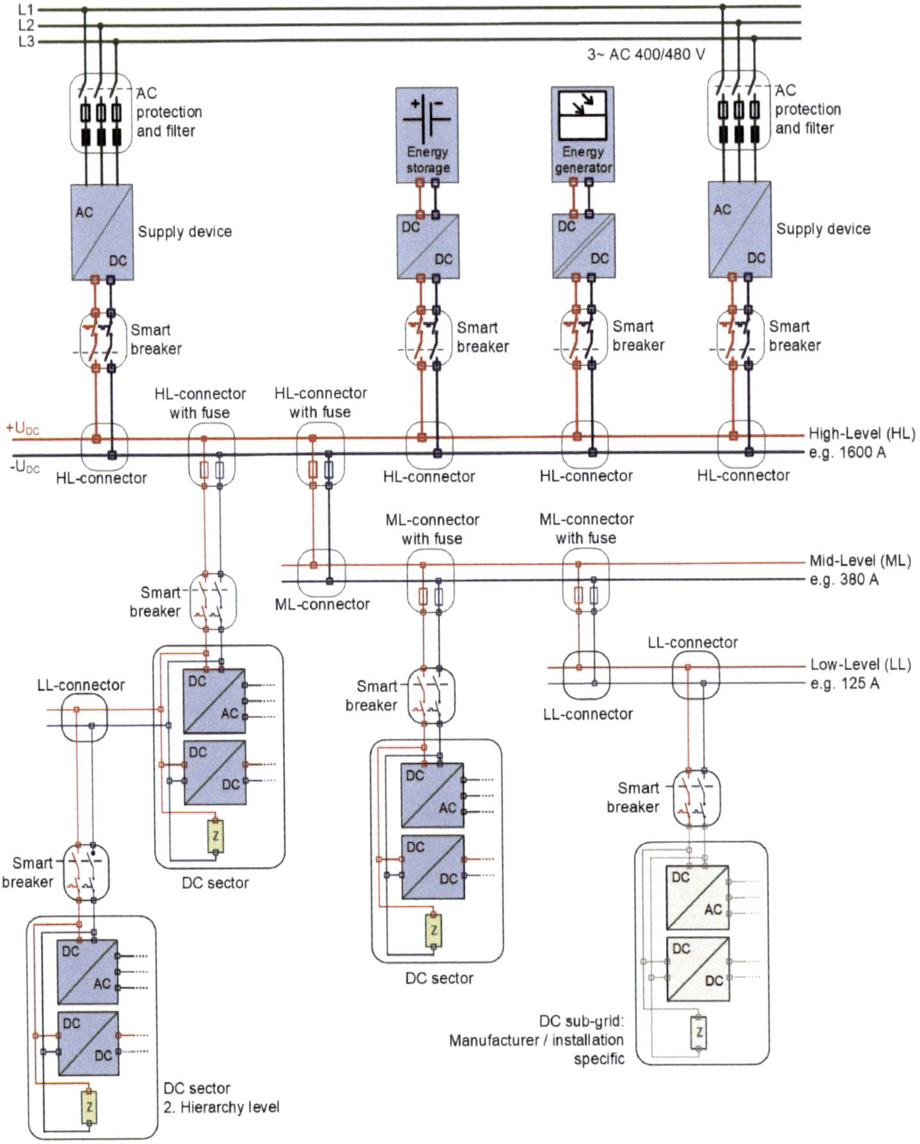


**Figure 5.1** Structure of the industrial DC grid

Basic principles, typical grid participants, as well as relevant functions and features in the design of a DC grid are described in the following.

### DC Sectors

The basic principle of an open DC grid is that all participants are assigned to so-called DC sectors.

A DC sector is defined as a group of several DC devices that are directly connected to each other by the power flow and that are switched and protected together. They are present on every hierarchy level. All devices within a DC sector constitute a logical and functional unit. They are precharged or discharged together and disconnected from the DC grid together in the event of a fault. When connecting, both the power requirement and the spatial arrangement of the DC sector must be taken into account.

The following rules must be observed:

- Each DC sector is connected to the main grid via at least one Smart Breaker. The entire DC sector can be connected to the DC grid or disconnected from the DC grid via this Smart Breaker. The Smart Breaker is responsible for disconnection, protection and precharging of its DC sector. A detailed description of the functional tasks and the technical implementation of the Smart Breaker can be found in Section 5.5 (Smart Breaker for disconnection and protection) and Chapter 7 (Safety in the DC grid).
- Proprietary subgrids are permitted as DC sectors or within a DC sector. In this case, an interconnection of manufacturer-specific devices is connected to the DC grid, whereby all specifications of DC-INDUSTRIE (e.g. voltage ranges, EMC, short-circuit resistance, ...) would have to be adhered to at the connection point.
- DC sectors may disconnect from the DC grid and, if their own storage units or generators are available, they could continue to operate autonomously.
- Cascading of DC sectors is permitted.
- The interconnected devices of a DC sector are assigned to useful sub-functions of a machine/plant. A short-circuit fault within a composite device can lead to a loss of function of other composite devices of the same DC sector but would not affect the continued operation of other DC sectors (selectivity and protection by the Smart Breaker).
- With respect to dimensioning the power of the DC sectors, it is preferable to use - if possible - the same total rated power in each one. In a DC sector, the smallest participant should not deviate from the largest one by more than a factor of ten.

### Participants in a DC Grid

Within the DC sectors, there are different participants, which are described in more detail below.

1. Supply devices (AC/DC converters)

   Supply devices such as AC/DC converters establish a uni- or bidirectional connection between the AC grid and the DC grid. They always form a separate DC sector, which is commonly referred to as a grid feed or feed point. They generally feed the highest level of the DC system (DC main grid). In Figure 5.1, two identical grid feeds are sketched (AC/DC supply units top left and top right). In principle, any number of such feed points is permissible. Each supply point has its own AC breaker with the functions switching, overcurrent protection and EMC filter. Via the assigned Smart Breaker, the supply devices can charge the DC grid or be charged from the DC grid. Typical supply devices include inverters such as Active Infeed Converters (AIC) with power control, rectifiers with downstream boost/boost converter or simple bridge rectifiers with unregulated DC voltage. These technologies cannot be combined with each other arbitrarily.

2. Energy storage (via DC/DC converters or directly on the DC mains)

   Central storage units of higher capacity (e.g., accumulators or capacitor banks) are typically connected to the main DC grid. The connection is made via the assigned Smart Breaker (see Section 5.5) and a bidirectional DC/DC converter. Decoupling via a DC/DC converter allows the storage devices to be charged and discharged selectively, regardless of the current level of voltage in the DC mains.

   Capacitors can also be used without a DC/DC converter as fast, non-controllable energy storage devices to stabilize the DC voltage. These are then either connected centrally as a separate DC sector to the main DC grid via a Smart Breaker or used in a decentralized way within a DC sector to increase the capacity of the interconnected participants located there. In dynamic applications in particular, this buffer capacity can be used to absorb ripple currents that occur within the DC sector in a way that they are not transmitted "outside" to the higher-level DC grid. In principle, capacitors may be connected directly to the DC grid without an EMC filter. In contrast to switched-mode participants such as drive inverters, capacitors do not emit any EMC interference that would have to be limited.

3. Local energy generators (via DC/DC converters)

   In the DC grid, it is very easy to use local energy generators (e.g., solar panels or fuel cells) that feed in energy independently of the AC grid. It makes sense to connect them to the main DC grid at the same voltage level as the connection to the AC grid. Viewed from the source, unidirectional DC/DC converters should be provided initially for the connection then for the Smart Breakers. Depending on the voltage level, voltage class and grounding system, the DC/DC converters are to be designed as insulating converters.

4. Electric drives (with DC/AC converters)

   In the DC grid there are only variable speed drives. Even if this function is not required by the process, it can be used to advantage during start-up (soft start) and shutdown (guided braking) of the system.

   Instead of the otherwise widely used frequency converters with AC mains input and the switching devices of mains motors, simple inverters with DC intermediate circuit (symbolized as DC/AC converters in Figure 5.1) are used here. In principle, control cabinet devices, field devices or integrated drives (inverter motors) can be used to implement these variable-speed drives.

   Electric drives such as flywheel mass storage units are a special case. Here, the rotational mass of a flywheel firmly coupled to the motor shaft is used as a fast bidirectional energy store.

5. Sockets for AC devices (via DC/AC converters)

   At least in the transition period, DC/AC converters are needed to generate a common AC voltage (as an island grid) at local points in a factory. Via CEE (3~400 V) or "Schuko"-sockets (protective contact socket, 1~230 V), this is used to supply consumers that are not yet available in DC technology; for example, mobile welding equipment, drilling machines or measuring devices.

6. 24 V auxiliary voltage (via DC/DC converter)

   Unidirectional DC/DC converters are used to convert the "high" operating voltage in the DC grid to the extra-low voltage required in automation technology. This can either be done at a central point or locally distributed. Today, 24 V is almost exclusively required as the auxiliary voltage supply; in the future, however, the values may also be smaller or larger.

   A special feature is the provision of auxiliary voltage during start-up of the DC grid. In order for the switching devices and monitoring equipment to be able to operate, they must already have their supply voltage before the DC grid is applied in all DC sectors. This can be ensured, for example, via 24 V AC/DC power supplies or battery storage.

7. Welding case (with DC/AC converter and transformer)

   Welding operations require very high power for very short periods of time. This is especially true for aluminum spot welding. Here, one welding process alone requires over 100 kVA for about 200 ms. This power can be provided by locally allocated rapid storage. In an extended grid, the energy should be supplied by the "main grid". It is assumed that the individual welding operations are statistically distributed in a factory building. If this is not the case, selective timing must be used. As with electric drives, the primary voltage for the welding transformer is provided by inverters (DC/AC converters) which are clocked at 1 … 20 kHz.

8. Ohmic, inductive or capacitive loads (via switches)

   In most plants there are energy consumers that are switched on or off depending on the process and that are directly connected to the mains. These include, for example, the entire lighting technology as well as heaters for various process media (adhesives, soldering systems, heating furnaces, ...).

9. Manufacturer-specific subgrids (via DC/DC converters)

   In the open DC grid, proprietary subgrids of a different specification (manufacturer-specific DC interconnection systems) and/or other voltage (e.g., DC 380 V) can also be connected via a crosspoint switch. In terms of type, these subgrids again include in principle all the interconnection participants mentioned so far.

### Levels of Different Current Carrying Capacity

To connect the different levels with each other, there are dedicated connectors (feeder boxes). On one level there can be several such outgoing feeders to different levels below. However, the lower levels must not be connected to each other on the DC side (avoidance of loops).

When changing the cross-section, the line protection must be observed. This can be provided by circuit breakers with isolating characteristics or DC fuses.

### Device Classes

If a device within a DC sector has a fault, it cannot be ruled out that this fault will also affect other devices within the DC sector. There are two classes of devices with regard to self-protection:

Class 1 devices can potentially fail in the event of a short circuit at their DC terminals to the extent that an internal protection device (e.g., a fuse) trips. This internal protective device is an effective means of disconnecting the device from the DC mains in the event of an internal fault, thus preventing a fire (UL approval).

Class 2 devices ensure that they are not defective in the event of a short circuit at their DC terminals, i.e., that they are ready to be switched on again without further action once the fault has been rectified. If the manufacturer uses a fuse for fire protection in the event of an internal fault (short-circuit/carth fault due to component defect, conductor path short-circuits or similar), this fuse must not blow in the event of a short-circuit at the DC terminals. Fire protection in the event of such a device-internal fault can also be ensured by means other than a fuse, e.g., by the appropriate dimensioning of terminals, conductor tracks and components.

The energy fed back into the DC grid in the event of a short circuit corresponds to the amount of energy stored in the device-internal DC link capacitance. For drives, the energy of a connected synchronous machine (incl. load), which depends on the operating point, can also be added. The pole-wheel voltage proportional to the speed drives the motor short-circuit current via the inverse diodes in the inverter into the fault location until the machine stops.

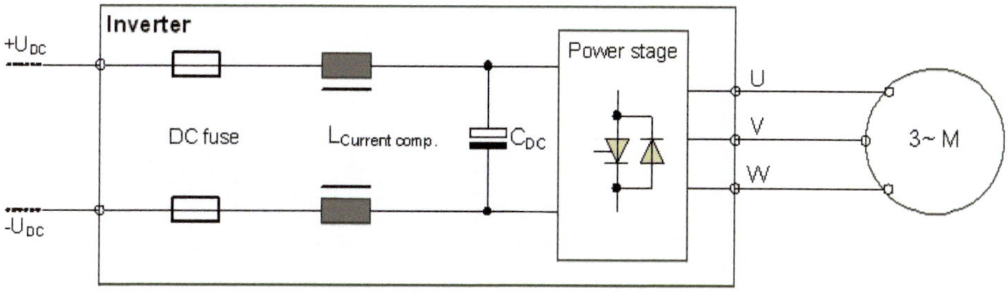


**Figure 5.2** Circuit topology for a device class 1 drive

Figure 5.2 shows the typical structure for a class 1 device. The DC fuses trip equally in both directions in the event of an overcurrent. Although this ensures self-protection in the event of an internal fault, the problem of a device failure arises as soon as a short circuit occurs in the neighboring device or in the DC grid. Typically, class 1 devices are therefore operated as individual devices in their own DC sector, as is the case with power feeders, DC/DC converters for storage/generators or main drives. Care must be taken to ensure that the circuit breaker in the associated Smart Breaker trips faster than the internal fuse.

## 5.2 Parallel Operation of Supply Units

Different participants work together in the DC grid. To ensure parallel operation of supply devices and storage units without superimposed communication, the concept of characteristic-based control has been further developed. This is a simple procedure to control the power flows. The procedure must be implemented in all supply and energy storage devices. All devices are current controlled and the setpoint of the current is determined with a characteristic curve from the actual DC mains voltage $u_{DC}$. Figure 5.3 shows exemplary characteristic curves of three devices in parallel operation. For the quantities $u_{DC}$ and $i_{DC}$ the consumer counting arrow system[1] applies, which is shown in Figure 5.4.

[1] Negative sign for the current/electricity when a producer delivers power.

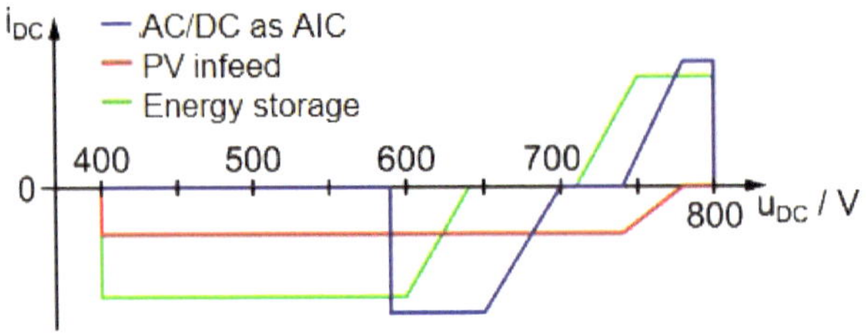


**Figure 5.3**
Exemplary control characteristics of different grid participants with active power control

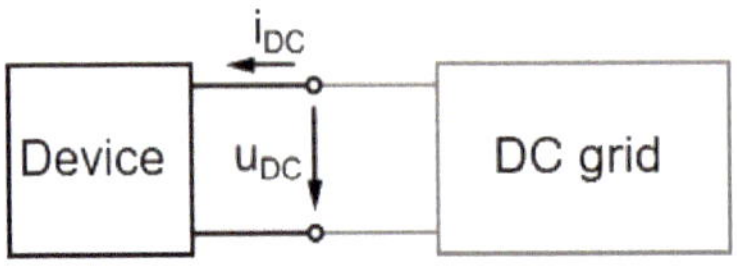


**Figure 5.4**
Definition of $\mathbf{u}_{DC}$ and $\mathbf{i}_{DC}$

The AC/DC converter as AIC (Active Infeed Converter = blue control characteristic) can only feed energy into the DC grid if the DC voltage is higher than the rectification value of the applied AC grid voltage. In the Figure, a 400 V AC grid is assumed, so the utility operates at full power from $u_{DC}$ ≥565 V. At higher DC voltage, it reduces its feed-in power down to 0, finally feeding back into the AC grid from about DC 750 V.

The accumulator (energy storage = green control characteristic) is inactive in the "nominal voltage range" of the DC grid. When the DC voltage drops, it releases energy into the DC grid, while when the DC voltage is too high, it counteracts the rise by absorbing energy.

The PV feed-in (solar system via DC/DC converter = red control characteristic) can only feed energy into the DC grid. This takes place almost in the entire control range of the DC grid. Only in the case of extremely high DC voltage this feed-in has to be regulated so that the maximum value of DC 800 V in the DC grid is not exceeded.

The following sections specify the implementation and parameterization of the control characteristics in the devices. In addition, special features of the parallel connection of "energy sources" as well as the integration of devices with limited control capability are discussed.

### Droop Curve for Supply Devices

The droop curves for supply devices define the relationship between $u_{DC}$ and $i_{DC}$. $u_{DC}$ is the DC mains voltage at the terminals of the device, see Figure 5.4. The load-count arrow system applies, i.e., the current $i_{DC}$ is positive when the current flows into the device from the DC mains (sink mode) and negative when it flows out of the device (source mode).

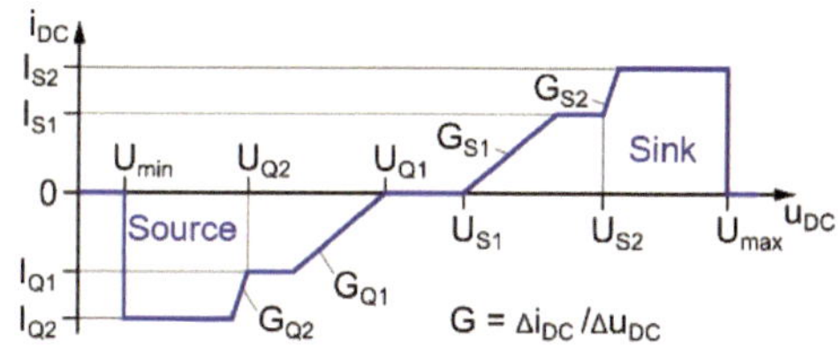


**Figure 5.5**
Definition of the droop curve

The droop curve is designed as shown in Figure 5.5 and defined by 14 parameters (Table 5.1).

**Table 5.1** Parameters of the Droop Curve

| Parameter | Description | Unit | Range of values |
|---|---|---|---|
| $I_{Q2}$ (I_Q2) | Maximum (negative) current in source mode | A | $I_{Q2} \leq I_{Q1}$ |
| $I_{Q1}$ (I_Q1) | Current of the plateau in source mode | A | $I_{Q1} \leq 0$ |
| $I_{S1}$ (I_S1) | Current of the plateau in sink mode | A | $0 \leq I_{S1}$ |
| $I_{S2}$ (I_S2) | Maximum current in sink mode | A | $I_{S1} \leq I_{S2}$ |
| $U_{min}$ (U_min) | Minimum voltage | V | $400\ V \leq U_{min} < U_{max}$ |
| $U_{Q2}$ (U_Q1)<br>$U_{Q1}$ (U_Q2) | Voltage of the 1st/2nd kink point in source mode | V | $U_{Q2} < U_{Q1} \leq U_{S1}$<br>$U_{min} \leq U_{Q2} < U_{Q1}$ |
| $U_{S1}$ (U_S1)<br>$U_{S2}$ (U_S2) | Voltage of the 1st/2nd kink point in sink mode | V | $U_{Q1} \leq U_{S1} < U_{S2}$<br>$U_{S1} < U_{S2} \leq U_{max}$ |
| $U_{max}$ (U_max) | Maximum voltage | V | $U_{min} < U_{max} \leq 800\ V$ |
| $G_{Q2}$ (G_Q1)<br>$G_{Q1}$ (G_Q1) | Gradiant of droop curve section 1/2 in source mode | A/V | $G_{Q1} > 0$<br>$G_{Q2} > 0$ |
| $G_{S1}$ (G_S1)<br>$G_{S2}$ (G_S2) | Gradiant of droop curves section 1/2 in sink mode | A/V | $G_{S1} > 0$<br>$G_{S2} > 0$ |

In the droop curve, the current $i_{DC} = 0$ in the following voltage ranges:

$u_{DC} \leq U_{min}$; $U_{Q1} \leq u_{DC} \leq U_{S1}$ and $u_{DC} \geq U_{max}$.

A vertical shift of the droop curve sections is achieved by shifting the kink points. The gradiants of the droop curve sections can be parameterized directly. The kink points should be in the normal operating range of the DC grid or in the steady-state overvoltage or undervoltage range. For $U_{Q1}$, the rated voltage (650 V) should be set as the factory setting for supply units and the set gradients should ensure stable behavior.

If the plateaus at $I_{Q1}$ and $I_{S1}$ are not desired, $U_{Q2} = U_{Q1} + I_{Q1}/G_{Q1}$ and $U_{S2} = U_{S1} + I_{S1}/G_{S1}$ are selected and the characteristic shown in Figure 5.6 is obtained without plateaus.

If the kink points at $I_{Q1}$ and $I_{S1}$ are not desired, $I_{S1} = I_{S2}$; $I_{Q1} = I_{Q2}$; $U_{min} = U_{Q2}$ and $U_{S2} = U_{max}$ are selected. This results in the droop curve shown in Figure 5.7.

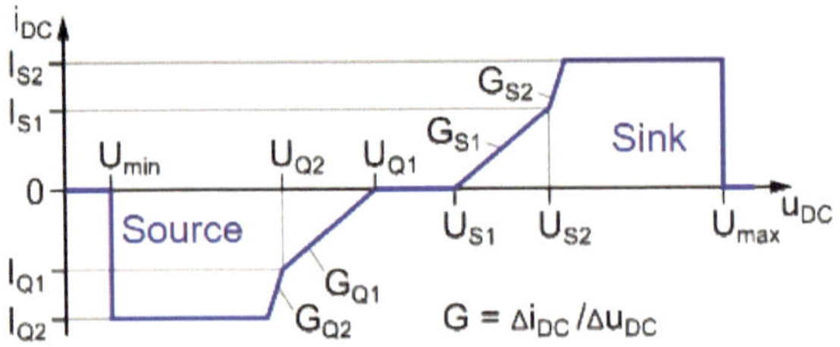


**Figure 5.6** Droop curve without plateaus

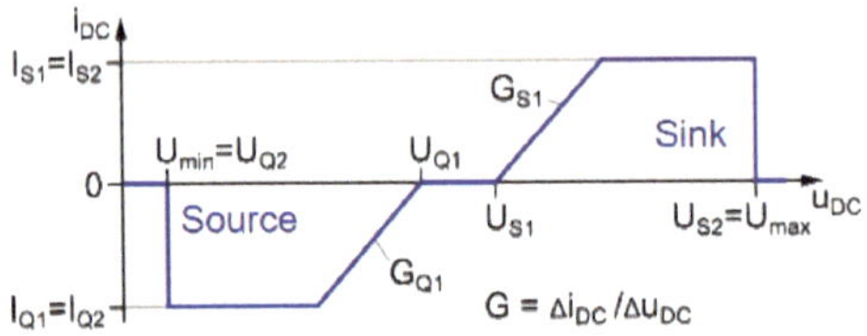


**Figure 5.7** Droop curve without plateaus

The control system of the group control to be implemented in the devices is shown in Figure 5.8. In order to influence the dynamics and to ensure stability, a first order lag element element is necessary at the input of the droop curve, which can be parameterized via the time constant $T_{FU}$. To prevent erratic changes of $i_{DC}$ when the droop curve changes, another first order lag element at the output is optionally possible (time constant $T_{FI}$). The parameters of the filters are listed in Table 5.2.

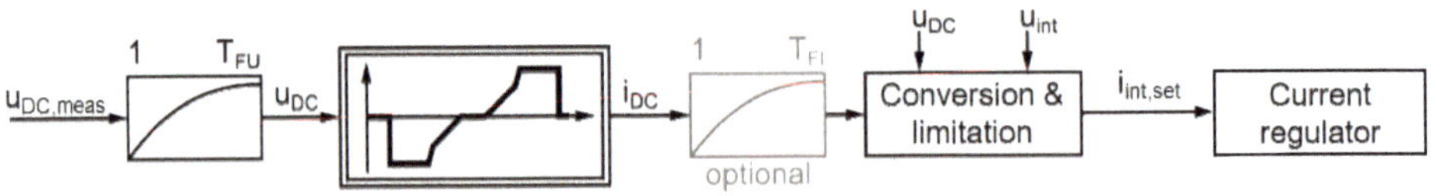


**Figure 5.8** Group control system

**Table 5.2** Parameters of the Group Control Filters

| Parameter | Description | Unit | Range of values |
|---|---|---|---|
| $T_{FU}$ (T_FU) | Time constant of the filter of $u_{DC,meas}$ (low pass first order) | s | $T_{FU} \geq 0$ |
| $T_{FI}$ (T_FI) | Time constant of the filter of $i_{DC}$ (low pass first order), This filter is optional | s | $T_{FI} \geq 0$ |

Depending on the type of device, the setpoint for $i_{DC}$ must be converted into an internal current setpoint $i_{int,set}$ so that the power $p_{DC} = u_{DC} - i_{DC}$ is set. A dynamic limitation of this setpoint must be implemented individually depending on the device state and the overload capability.

If an overload persists, the device may limit the current to a lower value (e.g., the rated current) and issue a warning, but not go into a fault condition and shut down.

### Parallel Connection of Feeders

When connecting feeders in parallel, symmetrical division can be easily parameterized by defining the gradients. The voltages of the bend points $U_{Q1}$, ... are set to the same values. The gradients ($G_{Q1A}$ for feeder A and $G_{Q1B}$ for feeder B) must be selected in the ratio of the rated currents ($I_A$ for feeder A and $I_B$ for feeder B) as follows:

$G_{Q1A}/G_{Q1B} = I_A/I_B$.

This results in the symmetrical current distribution shown in Figure 5.9 and the rated currents of the two feeders are reached at the same voltage.

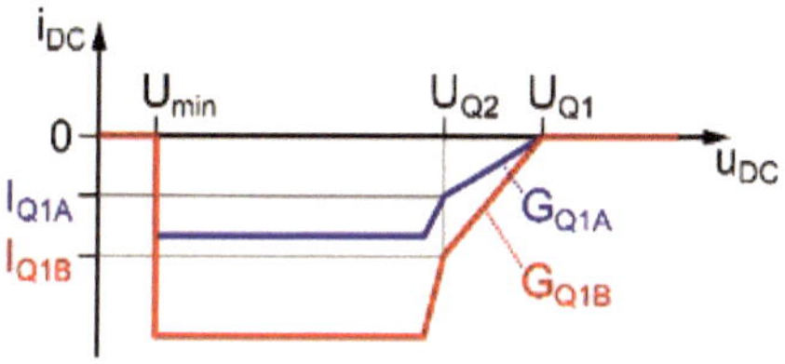


**Figure 5.9**
Control characteristics for parallel connection of power feeders

### Operating Area for Variable Power Feeders

If the feed-in device can only feed-in a limited power that depends on external factors, a permitted operating area must be defined. This is the case, for example, for renewable energy feeders. The permitted operating area can be seen in Figure 5.10. The device can feed-in any current within its capabilities. However, depending on the mains voltage, the maximum feed-in current is limited when the mains voltage comes close to the kink points $U_{Q1}$ and $U_{Q2}$.

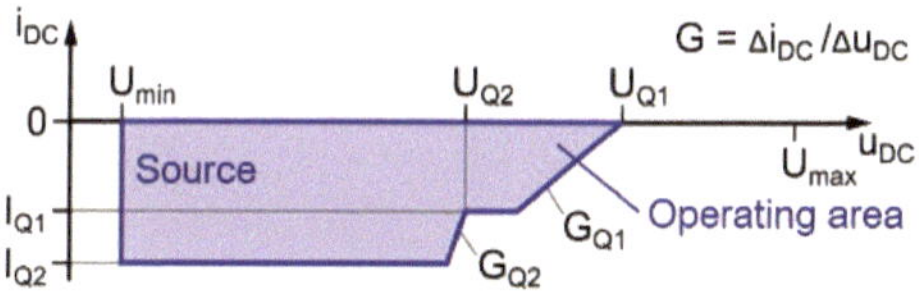


**Figure 5.10**
Operating area of a feed-in device of variable power (e.g. from renewable energies)

### Devices with and without Load Reduction

All devices in the DC grid may only draw or deliver current within defined voltage limits ($U_{min} < u_{DC} < U_{max}$). This operating area is shown in Figure 5.11. When exceeding or falling below the limit, the device must switch off.

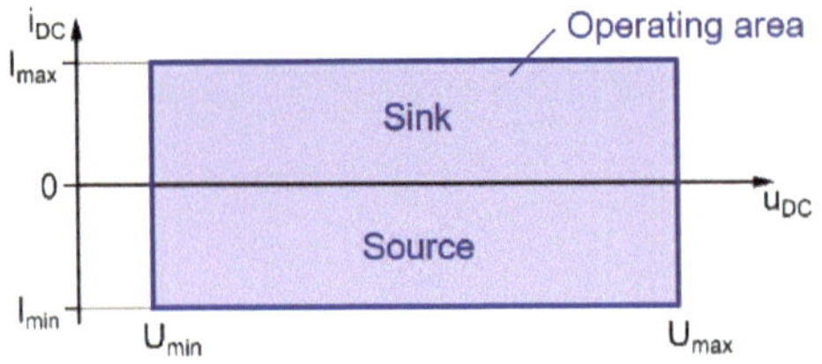


**Figure 5.11**
Operating area of a device without load reduction

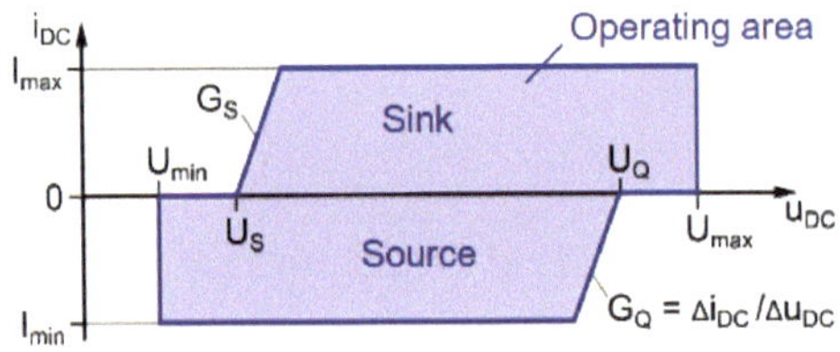


**Figure 5.12**
Operating area of a device with load reduction

If the device supports continuous load reduction (e.g., fan drives), the operating area is limited, as shown in Figure 5.12. At low mains voltage, this limits the maximum current drawn. At high mains voltage, the maximum delivered current is limited. It is also possible to switch-off only if the gradiants are selected very large.

## ■ 5.3 DC Voltage Band and Operating Behavior

Two permissible voltage bands are defined for the DC grid, the value ranges of which are based on the AC grids used worldwide. The definition of voltage bands is necessary because the DC grid is usually not operated with a fixed voltage for regulated feeders and energy storage systems. The decentralized approach with control characteristics allows the grid voltage to move in defined voltage bands. The grid voltage represents the current power state of the grid and defines the distribution of the power to the supply devices.

The respective rated operating range is obtained by rectifying 3~ 400 V +10/-15% or 3~ 480 V +10/-15% taking into account an unloaded mains. The following applies: $U_{DC} = \sqrt{2} * U_{AC,eff}$. In order to use the highest possible DC voltage in the regulated DC grid, the operating range for this type of supply considers the level of the unregulated 480 V AC grid. The consideration of these voltages has the advantage that current devices and installations are suitable for use in the open DC grid without fundamental modifications.

All components in the DC grid should not become defective in the voltage range from DC 400 to DC 800 V. Depending on the feed-in, however, they do not have to be operable in the entire range. The voltage ranges are shown in Figure 5.13.

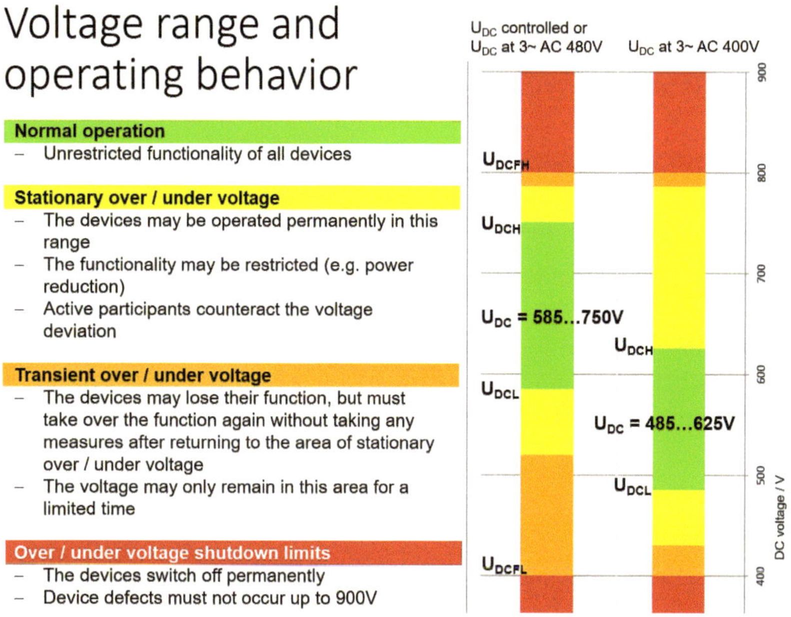


**Figure 5.13** Definition of the voltage bands in the DC grid

## Normal Operating Range

The normal operating range is characterized by a power balance between local generation (e.g., from PV systems), feed-in from the AC grid and consumption of the loads. All equipment in the grid is designed to operate continuously in this range and must be able to operate in this voltage range at its specified rated power. Limitations may arise due to physical boundary conditions, such as the state of charge of storage units, the level of the grid voltage in the case of feeders from the AC grid, or the solar radiation in the case of PV systems. These constraints lead to a possible, temporary imbalance of the energy flows. Other supplying devices must be able to compensate for this imbalance in the normal operating range.

The grid control concept for load flow control is designed in such a way that the system is always kept in this operating range if possible. With reference to the control characteristics, it can be seen that in this operating range a PV system, for example, is always allowed to feed its currently maximum available power/current into the grid. The AC/DC interfaces increase their feed-in linearly with decreasing value of the grid voltage level until a maximum current of the feed-in is reached.

The upper limit of the normal operating range was chosen to be the voltage that would occur at 110% of the rated voltage of the AC mains under no-load conditions when operating with a passive diode rectifier. Today, it is not guaranteed that all devices developed for operation on 400 V AC mains can also be used without restrictions at the maximum voltage of the actively controlled DC mains. Particularly in the case of standard motors without special insulation for inverter operation, operation with very high DC voltages can lead to impairment of the service life due to partial discharge (see e.g., IEC/TS 60034-25). If necessary, corrective measures (also IEC/TS 60034-25) must be taken or the operating time must be restricted in the range of the maximum voltage.

### Stationary Overvoltage and Undervoltage

In the case of steady-state overvoltage, there is excess power in the DC grid; local generation exceeds consumption, and sources coupled to the AC grid no longer feed power into the grid but feed it back into the AC grid if possible. The load flow management system tries to counteract this. Possible measures include feeding power back into the AC grid, reducing local generation linearly or in steps, charging energy storage devices, or converting the excess power into heating circuits or braking resistors.

For the steady-state undervoltage band, the conditions are reversed. In this case, the local generation is very low compared to the load demand, while the AC grid-connected sources inject their maximum current to cover the consumption in the DC grid. Load flow management attempts to counteract this with appropriate measures. These measures include increasing the feed-in from the AC grid if possible, discharging energy storage devices, and/or prioritized step shedding or down regulation of loads (so-called "load shedding").

For both areas, all devices must be designed for continuous operation in this voltage band. However, it is permissible that equipment in this range can no longer operate at its rated power. The reduction of the deliverable power must be specified in the data sheet of each component.

### Overvoltage and Undervoltage Disconnection Limits

Regardless of the feeding AC grid and the technology used for the feed-in, a value of 800 V is defined as the overvoltage cut-off limit to protect the devices. This value for the overvoltage disconnection limit is based on the characteristics of the inverters used in an AC grid today. It allows the use of a series connection of two capacitors specified for 400 V. A value of 400 V is specified for the undervoltage disconnection limit. If a component is switched off by reaching one of these limits, a fault acknowledgement by the operating personnel of the system is permissible or required before switching on again (USA/Canada).

The voltage band ranges between the shutdown limits and the respective steady-state overvoltage or undervoltage range are referred to as transient overvoltage or undervoltage ranges. These ranges are characterized by the voltage being outside the specified operating range. The respective voltage band ranges are reserved only for brief transient events, e.g., start-up/switch-on of large loads or fault conditions such as load shedding or voltage collapse due to a short circuit prior to tripping of a protective element. All devices remain connected to the DC mains in this voltage range. They may reduce their functionality in this voltage range but must resume this independently and without delay after returning to the steady-state overvoltage or undervoltage range.

The actions taken by the load flow management system to return the voltage to the specified range are similar to those taken in the steady-state overvoltage and undervoltage range. In the overvoltage range, this means switching off local generation, switching off AC injection or maximizing regeneration into the AC grid, extending the maximum possible power into existing energy storage devices, switching on devices for converting electrical energy, e.g., into heat (braking resistor). In the undervoltage range, the measures of the load flow control system are the same accordingly. They include the extension of the maximum feed from the AC grid and from existing energy storage devices as well as the shedding of loads.

### Voltage Change Rate

Any change in the load results in a change in the voltage in the DC grid. The voltage swing and rate of change depend on the dynamic load requirements of the specific application, the capacitances in the system grid, and the setting of the control characteristics.

The rate of change of the DC voltage results in a current and thus a thermal load on the capacitors. Large voltage strokes can also lead to premature aging of electrolytic capacitors.

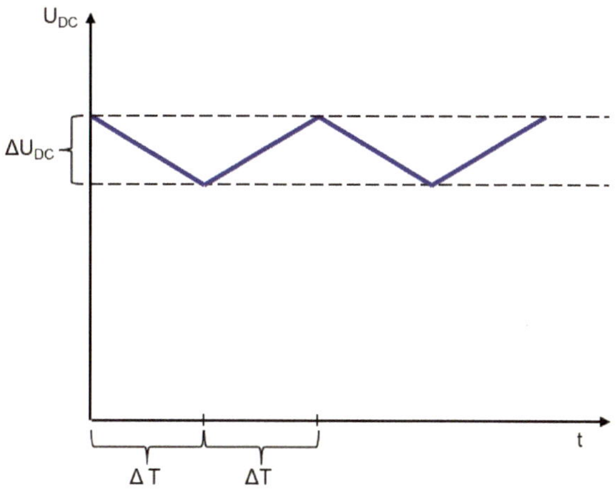


**Figure 5.14**
Course of the DC voltage during a cyclic load cycle

The load cycles for which the capacitors must be designed are described below. When designing the systems, care must be taken to ensure that the load cycles that occur do not lead to higher loads. Usually, critical load cycles only occur when there are dominant loads with large load fluctuations in the DC grid. Dominant means that the power is large compared to the total power of the grid. If necessary, the voltage swing must be reduced by additional capacitors or highly dynamic storage.

For the device design, it is assumed that the DC voltage in an extended DC grid changes with a maximum of $\Delta U_{DC}/\Delta T$ = *13 V/ms*. For a DC grid with a nominal voltage of DC 650 V, this corresponds to a 10% change in voltage in 5 ms. Figure 5.14 shows the course of the DC voltage during a load cycle with the maximum rate of change.

### Load due to Large Voltage Strokes

Common electrolytic capacitors can withstand a negative voltage swing of about 70 V without aging unexpectedly fast. Since two such capacitors are connected in series in the DC grid, we arrive at 140 V purely by calculation. With 10% safety because of unbalance, component tolerances, etc., this finally results in a value of 125 V. Voltage strokes greater than 125 V are permitted a maximum of 100 million times during the system service life. Otherwise, the electrolytic capacitors will age prematurely.

This means, for example, that a load cycle in which this voltage stroke is exceeded once each time may have a repetition frequency of max. 0.5 Hz if an operating life of 55,000 h is to be achieved.

If necessary, the voltage swing must be reduced by additional capacitors or highly dynamic accumulators.

### Protection against Overcharging

Protection against overcharging the DC grid is carried out on two different levels. First, in the voltage band ranges “steady-state” as well as “transient overvoltage”, active measures defined by the centralized or decentralized grid management are taken by the sources and by the consumers to prevent further deviation of the voltage from the normal operating range. This includes limiting feed-in from local renewable generators and AC/DC interfaces, increasing feed-in to storage systems and feeding back to the higher-level AC grid, and increasing consumption from flexibly controllable loads. This procedure is also the basis for the design of the control characteristics. The shutdown threshold (disconnection of all open-loop and closed-loop control circuits) for all devices is set to DC 800 V. Even without intervention of the grid management, no device may feed any more power into the DC grid above this threshold.

## ■ 5.4 Supply and Grounding Concept

The DC grid can be operated in different designs with regard to its connection to the AC grid. Initially, two concepts were selected and executed which allow the use of standard devices whose clearances and creepage distances are designed for operation on 3~400 V/480 V grids.

For smaller systems, operation is related to AC grids that are already present as a connection/installation in the factory building, and to which other AC consumers are also connected.

For extended DC grids, operation on a separate isolating transformer is favored, which provides isolation/decoupling from the AC grid. The ground potential of the system is thereby established in the DC grid via impedances. Such an isolated DC grid offers the advantage that only one of the two DC potentials ($+U_{DC}$ or $-U_{DC}$) in the entire installation has to be protected by means of fast circuit breakers/Smart Breakers. This reduces component and installation costs, but also the voltage drops in the DC grid.

### AC-Side Grounded Grid

The DC grid grounded on the AC side is operated on an AC grid in which the star point of the feeding transformer is grounded. The earth reference is specified via the AC grid. Other loads can be operated on this AC grid. In the DC grid, all Smart Breakers must be equipped with a 2-pole disconnector and a 2-pole circuit breaker. The supply devices must not shift the potential of the DC busbars, which is on average $+\frac{1}{2}\, U_{DC}$ or $-\frac{1}{2}\, U_{DC}$, at high frequency. This means that the active infeed converters with 2-level inverters usually used today for local DC grids cannot be used here. However, comparable solutions with common-mode filters are feasible.

Diode rectifiers are often used, as outlined in Figure 5.15. For the DC potentials, this results in a common-mode voltage of approx. 16% of the line voltage against PE with three times the line frequency, as with today's standard converters.

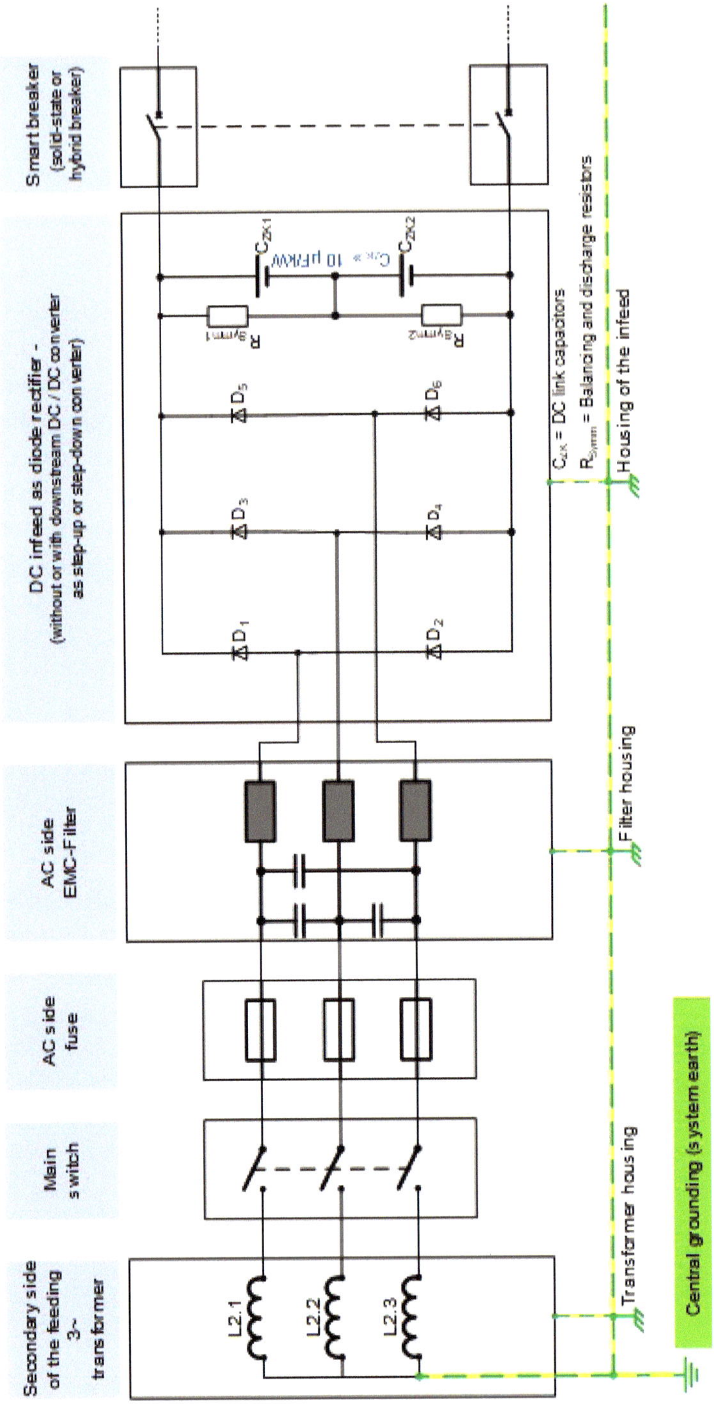


**Figure 5.15** DC grid with AC-side grounding

### DC IT Grid

The DC IT grid is operated on an AC grid where the star point of the feeding transformer is not grounded (Figure 5.16). The transformer is intended exclusively for connection to the DC grid. If further AC consumers are to be operated, they either require their own transformer or they are connected in the DC grid via DC/AC converters or they must be able to cope with the pulsed common-mode voltage. In the DC grid, it is sufficient if the Smart Breakers are designed with a 2-pole disconnector and a 1-pole circuit breaker. The PE potential is established via high-impedance resistors (assigned to the feed points) and Y capacitors (in the supply units and drive inverters) and connected to ground at a central point. This results in a balancing of the two DC potentials against earth (high-impedance midpoint grounding). In the event of an initial grounding fault, the system can continue to operate. This grounding fault must be detected and corrected within a reasonable time so that individual components are not permanently loaded with the higher voltage. If a second grounding fault occurs before the first has been rectified, the plant is shut down via a circuit breaker (the second grounding fault now causes a short-circuit between the two poles of the DC grid).

The term "grounding" pertains to two functions:

**Operational grounding** = Connection of DCP and DCN to PE via high-impedance resistors and Y-capacitors.

**Protective grounding** = The low-resistance connection of PE to the foundation earth electrode is made at a central point (system earth electrode). All conductive enclosures of equipment are connected to PE. Depending on the structure/anchoring, the housings may be additionally grounded.

The resistance of the operating ground must be greater than 75 kΩ so that in the event of a fault no more than 10 mA DC current flows to PE at the maximum operating voltage in rated operation of 750 V DC.

Each device provides capacitors to the PE. They are used for EMC interference suppression and, as a guide, must be a factor of ten greater than the parasitic leakage capacitances of the load connected to the device (lines, motors, transformers). Resistors are to be provided only in DC sectors with supply devices and storage units. Resistors must take into account that the ohmic grounding of the entire DC grid does not fall below at least 75 kΩ. The PE conductor is grounded with low resistance at a single point. There, the PE is connected to the foundation earth electrode. Due to the symmetrical design of the DC grid with respect to PE and the low leakage currents, no corrosion problem is expected at the foundation earth electrodes.

Due to the choice of grounding, the voltage load on the motors is not higher than previously for converter operation with diode rectifiers. It is ±2 × DC voltage for the conductor-to-conductor isolation and ±1.5 × DC voltage for the conductor-to-ground isolation. If a first grounding fault occurs on a DC busbar in the DC IT system, the voltage load of the conductor-earth insulation also increases to ±2 × DC voltage for the duration of this fault.

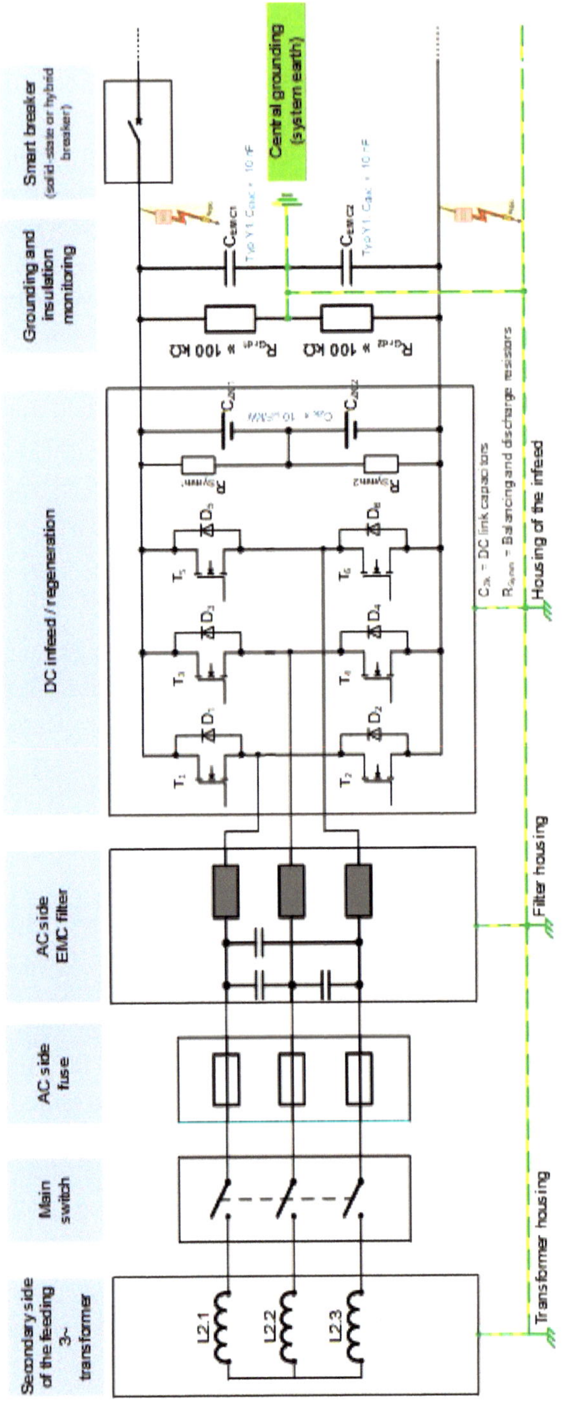


**Figure 5.16**
DC grid with high-impedance mid-point grounding (DC IT grid)

## ■ 5.5 Smart Breaker for Decoupling and Protection

The Smart Breaker decouples a DC sector from the next higher level, e.g., from the main DC grid; it contains all functions for switching on, operating and protecting the DC sector. The Smart Breaker optionally includes a mechanical disconnection point so that the DC sector can be safely disconnected from the main DC grid - for example - for maintenance work. The concept does not specify on which connection side of the Smart Breaker the mechanical disconnection point is located. Since generators or storage units can also be located within a DC sector, the generally applicable safety rules for working on electrical systems must be observed. Figure 5.17 shows a schematic design example for a Smart Breaker. Different requirements can be placed on a Smart Breaker at different hierarchy levels, which then results in topology or function variants of the Smart Breaker.

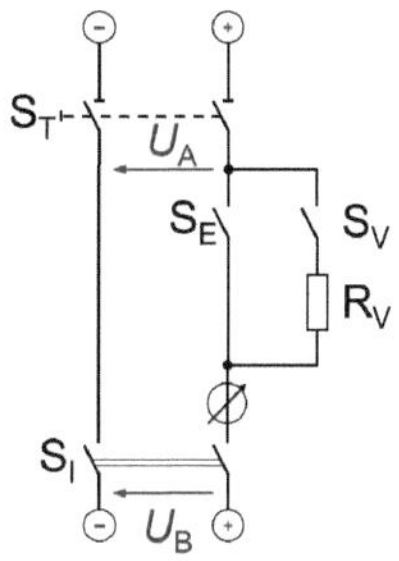


**Figure 5.17**
Example of execution for a Smart Breaker

The Smart Breaker includes the following basic functions, which are integrated optionally or obligatorily:

### Safe Disconnection via a Two-Pole Switch

The two-pole switch $S_T$ is used to disconnect a DC sector - for example - for maintenance work. It can be manual or motorized. Furthermore, it is optional. It can be omitted for small DC systems where maintenance work is not required in one DC sector while voltage is present in another.

### Overcurrent and Short-Circuit Protection

The overcurrent and short-circuit protection ensures that a fault in one DC sector does not lead to an impermissible influence on other DC sectors. For this purpose, a fast overcurrent shutdown is provided in the Smart Breaker for both device protection and line protection. This protective device can be designed as a purely electronic circuit breaker or as a hybrid circuit breaker.

### Precharging

DC sectors generally contain capacitors, e.g., in the intermediate circuits of the inverters. Precharging is required to limit the inrush current of the capacitors. Without such limitation, the inrush current can cause a protective element to trip. In the simplest case, this precharging is done by means of a precharging resistor $R_V$. Precharging can be bidirectional via the Smart Breaker, i.e., a DC sector can be charged from the DC grid (e.g., if only loads are located in the DC sector), or the DC grid can be charged from the DC sector (e.g., if there are storage units or generators in the DC sector). Precharging includes voltage measurement on both sides of the Smart Breaker. To protect against thermal overload of the precharging resistor, the charging process must be monitored. If the maximum loss energy is exceeded, $R_V$ can be protected with the $S_V$ switch. If the Smart Breaker connects a hierarchy level of the DC grid in which there are no capacities on the input side, the precharging function is not necessary. This also applies in the special case that precharging is implemented centrally (controlled ramp-up of the DC voltage via the supply device(s)).

Precharging is performed autonomously and cascaded across the grid levels. The process is defined by voltage thresholds and waiting times and is described in Section 5.6.

### Protection against Overvoltage or Undervoltage

If the voltage of the DC grid exceeds the maximum value of DC 800 V (see Section 5.3 for definition of voltage bands), the DC sector must be disconnected from the DC grid. For voltage detection, the voltage measurement available for precharging can be used. The overcurrent circuit breaker is triggered as the switching element.

If the voltage in the DC grid falls below the transient undervoltage limit of DC 400 V, the DC sector must be disconnected from the DC grid. This is necessary because the precharge procedure must be re-triggered for voltages below DC 400 V. The overcurrent circuit breaker is triggered as the switching element.

### Discharge

Conceptually, the discharge of the DC sectors is done by the devices connected in the DC sector itself. The devices must ensure that internal loads, such as balancing resistors, discharge their capacitances within five minutes. The precharging resistor $R_V$ can also be used for a rapid discharge of the DC sector. However, this is not conceptually mandatory.

### Ground Fault Detection

The grounding concept for the IT system provides that the DC system can continue to operate in the event of a single-pole ground fault. Any ground fault must be reliably detected. The ground fault is detected by means of insulation monitoring, e.g., with the aid of voltage measurements from DC+ to PE and DC– to PE. Optionally, this monitoring can be performed in the Smart Breaker. The essential requirement is the location of the ground fault. In the simplest case, the circuit breaker of the associated DC sector trips (by discharging all Y-capacitors); otherwise, special electronics for ground fault location are required, which isolate the ground fault by impressing known test signals.

### DC Fuses

Fuses or conventional DC circuit breakers can be used as so-called disaster protection. These are designed to trip only if one of the other protective devices in the Smart Breaker fails. Irrespective of this, these are intended as the sole protective device for line protection (e.g., in the event of a cross-section tapering).

## ■ 5.6 Precharging and Unloading Concept

Precharging takes place on several levels. Firstly, the 1$^{st}$ level is precharged – the feeders and, if applicable, the storage units – and thereby the unloaded DC grid. All DC sectors are separated. The 2$^{nd}$ level with the DC sectors is precharged (DC grid/1$^{st}$ hierarchy level) with a time delay and triggered by the voltage increase of the DC grid voltage with a waiting time. Figure 5.18 shows a simplified view of the grid with the levels. If further DC sectors are connected below a DC sector (further hierarchy levels in the DC grid), their precharging (3$^{rd}$ level) is again staggered. Staggered precharging should take a maximum of ten seconds per level.

The Smart Breakers included in Figure 5.18 are shown in a simplified form in Figure 5.17. They optionally contain, if required – e.g., for maintenance purposes – a two-pole disconnector $S_T$, which can also be designed as a manual switch. In the following considerations, this disconnector is always regarded as already closed, unless otherwise stated. For precharging the subsequent DC grid or the assigned DC sectors, the precharge switch $S_V$ is closed. This starts the precharging via $R_V$. When precharging is completed, the precharging resistor is bridged via the electronic switch $S_E$ and then $S_V$ is opened. Alternatively, $S_E$ can be designed as a hybrid switch.

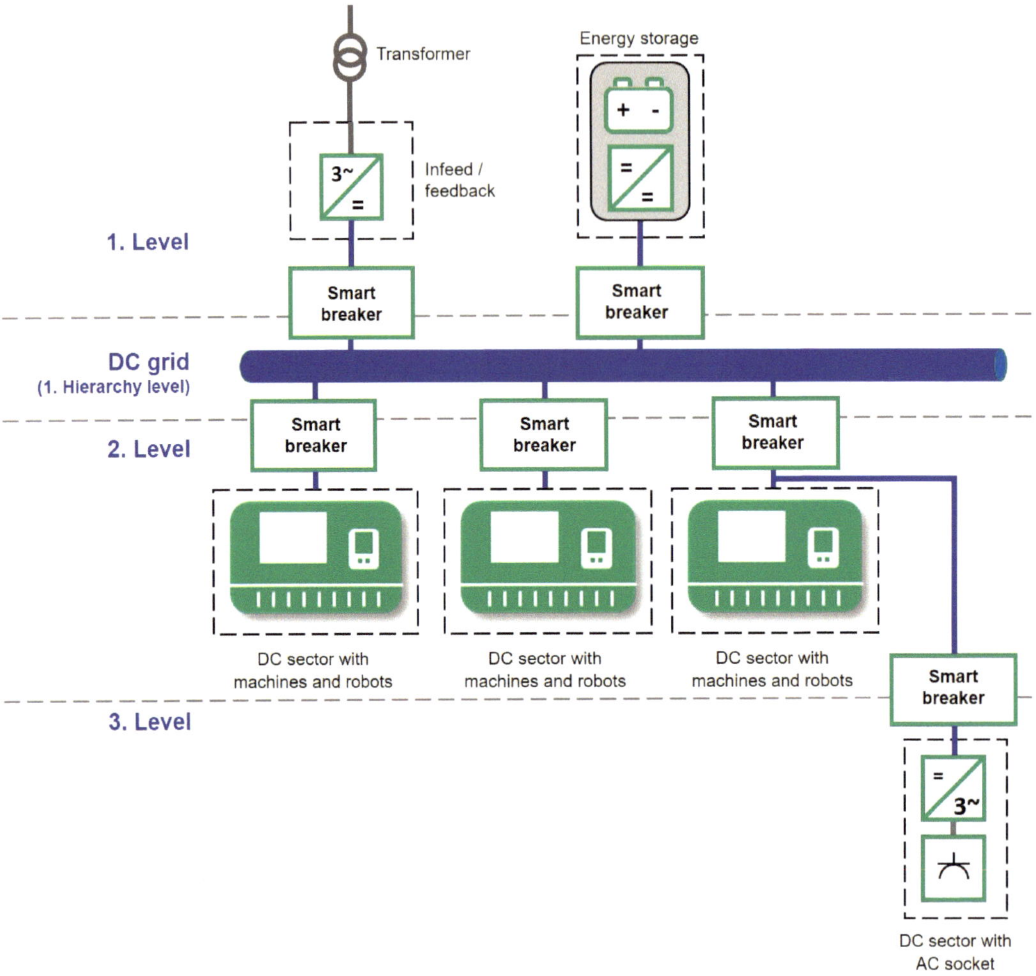


**Figure 5.18** Overview of the precharging levels

Different implementation variants are possible for the Smart Breaker:

a) Precharging via $R_V$. The resistor $R_V$ and the DC contactor $S_V$ are assigned to the switching and protection device and are controlled by it.

   Alternatively, precharging can be performed via a buck converter. The advantage of this is that fewer losses occur, and even very large capacitances can be charged with limited current. The realization can be either …

b) by an additional IGBT $S_V$ and the choke $L_V$ or …

c) using the already existing IGBT $S_{E1}$ and a combination of choke $L_V$ with bypass switch $S_{Ü}$.

These three variants are shown in Figure 5.19.

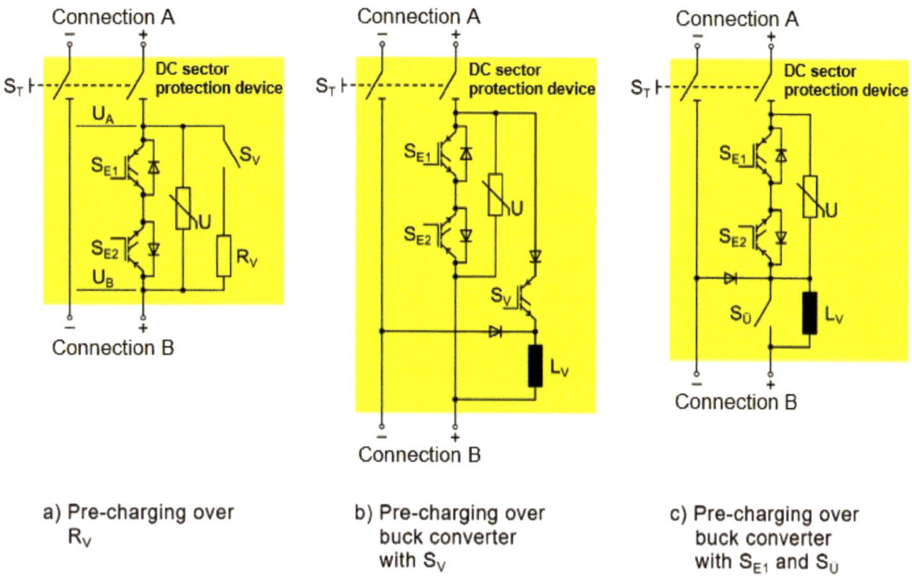


**Figure 5.19** Variants of the Smart Breakers for precharging

In principle, the precharging of the DC grid can be performed by a higher-level controller. In this case, the controller determines a sensible switch-on sequence for the switches and the individual precharging and bypass switches of the DC sectors are controlled directly.

However, it is expected that a DC grid without supervisory control will also exist. Therefore, a self-sufficient precharging concept is described below, where the logic is stored in the Smart Breaker and can run without communication with the grid management. It represents the minimum functionality and can be extended or replaced by the grid management. The precharging concept is only explained with respect to variant a). However, variants b) and c) can also be used in the DC sectors. Instead of simply closing $S_V$, in this case a current should be set in $L_V$ in a clocked manner.

## Design of the Precharge

In case of precharging via resistor $R_V$ (variant a)), the maximum current of precharging via $R_V$ is limited. The relative charging current is defined as a function of the rated current $I_N$ as $k_V = I_V/I_N$. For a specific capacitance of – for example – C' = 25 µF/kW in the DC sectors and a $k_V$ = 0.1, this results in a time constant $\tau = R_V \cdot C = U_N^2 \cdot C'/k_V$ = 106 ms. After 5 τ, the DC sector can be considered charged.

If loads such as auxiliary power supplies are already drawing power during the precharging of the DC grid or a DC sector, a steady-state voltage difference $\Delta U_B$ is established during the precharging process, which is balanced by closing $S_E$. As a result, equalizing currents flow which – in extreme cases – can cause the switching and protection device in the Smart Breaker to trip. They depend on the capacitances, their internal resistances, the inductive or resistive line loadings and the Smart Breaker.

The amount of load during precharging is defined as the relative load $k_B = P_B/P_N$ depending on the power of the load $P_B$ with respect to the rated power $P_N$. The voltage difference $\Delta U_B$ is approximately determined by $\Delta U_B = U_N \cdot k_B/k_V$. It thus depends on the relative load and relative charging power and is determined by measuring the voltage in the Smart Breakers on both sides. A maximum voltage difference $\Delta U_B = 15$ V is recommended as a limit. At $k_V = 10\%$, this corresponds to a relative load of approx. 0.2%.

During precharging of the DC grid, any auxiliary power supply units (AVU) of the Smart Breakers or DC sectors are effective, i.e., electrically connected prior to the Smart Breakers, as shown in variants 1 and 2 in Figure 5.20a. These must comply with this power limitation or – as described further down – delay their power consumption. The same applies to auxiliary power supplies within the DC sectors (variant 3 in Figure 5.20a).

For variants 1 and 2, it must be taken into account that faults in the auxiliary voltage supplies themselves are not switched off by the Smart Breaker of the associated DC sector and must therefore be controlled separately.

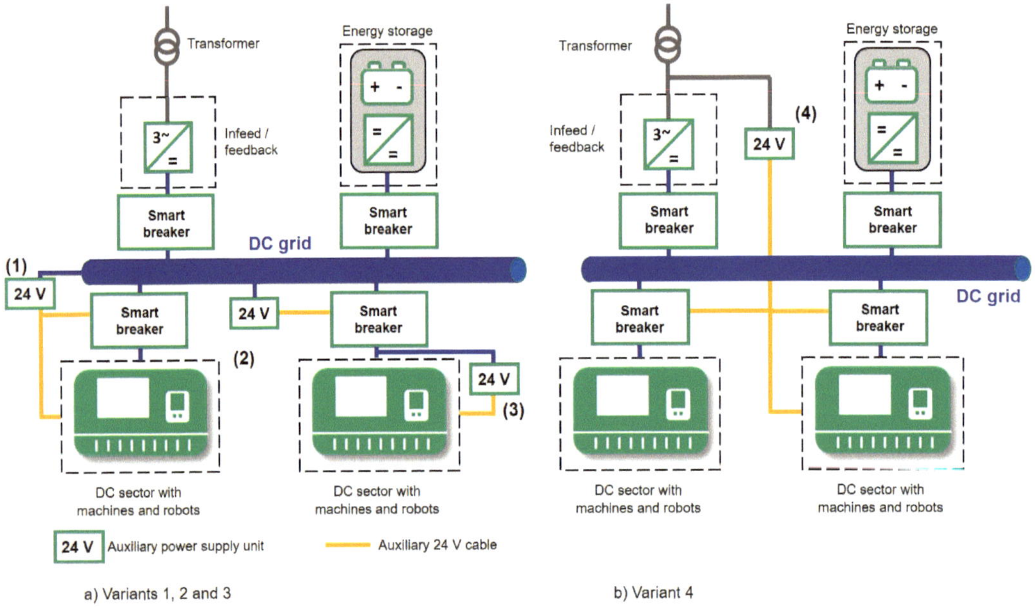


**Figure 5.20** Variants for the auxiliary power supply

If a large power of the auxiliary voltage supply is already necessary in the DC sectors during precharging or if the DC sectors have to be supplied before the DC sector is charged, this can be realized with a central auxiliary voltage supply unit. This is shown as variant 4 in Figure 5.20b.

**Precharging the First Feed-in Device from the AC Mains**

Before the first feeder is started, the DC mains is discharged and disconnected from the feeder ($S_V$ and $S_E$ open). The power supply unit precharges itself from the AC mains to reach the rectified mains voltage. It automatically switches to closed-loop operation after the internal charging current limit has been bypassed. This is started by setting the controller enable signal (RF). Owing to the stored characteristic curve, the rated voltage $U_N$ of the DC grid is set at the output point.

Connection A of the Smart Breaker is connected to the feeder (voltage $U_A$) and connection B is connected to the DC grid (voltage $U_B$), see Figure 5.18 and Figure 5.19. The precharging sequence is shown in Figure 5.21a. When $U_A \geq U_1$ is reached, time $T_1$ is waited. Within this time, the feeder must be in the controlled and loadable mode. For this purpose, its controller enable signal (RF) is issued in the startup process of the feeder. After $T_1$ has elapsed, $S_V$ is closed, which starts the precharging of the DC grid. All DC sectors are disconnected at this time. Only auxiliary power supplies supplied directly from the DC grid contribute to the capacity to be charged and can already load the grid. This load must be limited in its power or the auxiliary voltage supplies may only load the grid with a time delay (by $T_1$ after $U_1$ has been reached). The load causes a steady-state deviation $\Delta U_B$, which must be compensated by $S_E$ in a hard way. If the voltage difference $U_A - U_B \leq \Delta U$ is undercut within $T_2$, the switch $S_E$ is closed. Either directly after undercutting - this shortens the precharge sequence and is necessary at high load during precharge - or after the time $T_2$ has elapsed. In this latter case the voltage difference $\Delta U_B$ might therefore be lower. After $T_3$ has elapsed, the precharge switch $S_V$ is opened again and the DC grid can be loaded. If the voltage does not undercut the differential voltage or if an overtemperature of $R_V$ occurs in the meantime, the precharging process is aborted ($S_V = 0$) and an error is indicated. This would occur if $R_V$ is not present or if there is a short circuit in the DC sector. In the latter case, the switch $S_V$ must be able to switch off the current limited only by $R_V$.

The following values are suggested for the voltage limits and waiting times: $U_1$ = 520 V (see explanation in the following paragraph); $\Delta U$ = 15 V; $T_1$ = 5 s; $T_2 \leq 4$ s; $T_3$ = 1 s. As a boundary condition, $T_1 > T_2 + T_3$ must hold so that subsequent levels do not begin their precharge until the preceding level has completed its precharge. The time $T_1$ results from the time constant $\tau$ of the precharge. After the time 5 $\tau$, it can be considered fully charged. Therefore, $T_1 > 5\,\tau$ must be valid. It is advantageous if the voltage difference $\Delta U$ at the Smart Breaker can be adjusted to allow for operationally larger values.

The suggested voltage limit of $U_1$ = 520 V must be selected lower, depending on the type of feed(s) and system design, so that the supply unit can take over the supply again when the DC grid is in the range of steady-state or temporary undervoltage.

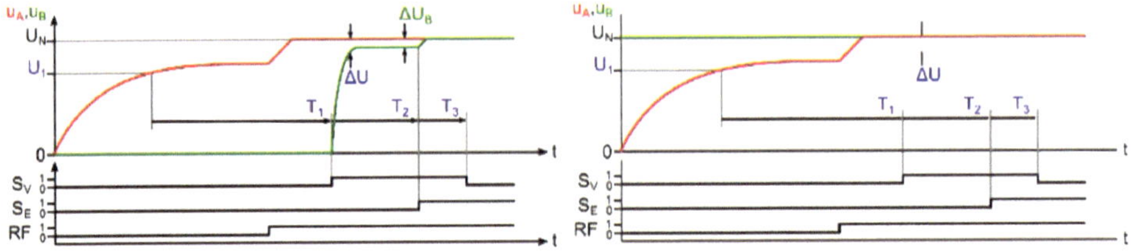


a) Pre-charging sequence of the first infeed device

b) Pre-charging sequence of each additional infeed device on the pre-charged unloaded grid

**Figure 5.21** Precharging sequences of the feeders (I)

If another feeder is started and the DC grid is unloaded, the precharging sequence is the same as for the first feeder with the difference that the DC grid voltage $U_B$ already corresponds to its nominal value at the beginning, see Figure 5.21b.

In case of an already loaded grid, the self-supporting precharging cannot be used without restrictions, because the actual grid voltage differs from the open-circuit voltage of the feeder which is to be connected. Both voltages are regulated, and the deviation may not be compensated by $R_V$. Depending on the load, therefore, it is aborted due to overload of the precharging resistor, see Figure 5.22a. If the load is low, the connection can still be completed successfully.

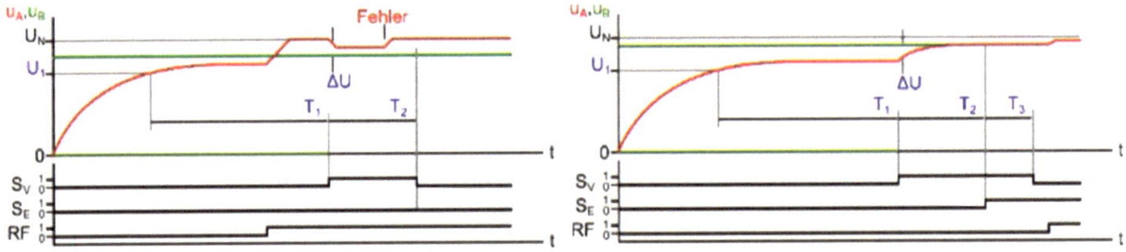


a) Pre-charging sequence of a further infeed device on the pre-charged and loaded grid with an error

b) Pre-charging sequence of an additional infeed device on the pre-charged and loaded grid with delayed controller release

**Figure 5.22** Precharging sequences of the feeders (II)

The solution of the problem is a delay of the controller release of the further infeed device. This is precharged from the AC mains only to the rectified AC mains voltage. When $U_A \geq U_1$ is reached, time $T_1$ is waited and then $S_V$ is closed. This results in the precharging of the feeder to the current DC grid voltage, regardless of the current load situation. The further sequence is as described above, except that only after the closing of $S_E$ and the expiration of $T_3$ the controller enable signal (RF) is given for the feed-in device and only then it participates in the load sharing, see Figure 5.22b.

### Precharging from Storage and other DC Voltage Sources

Precharging from storages and other DC voltage sources takes place in the same way as precharging from feeders. The prerequisite is that their respective output voltage exceeds the voltage $U_1$, that they do not emit any energy during the precharging process, and that they only switch to regulated operation after $T_2$ has elapsed.

### Precharging the DC Sectors

Here, connection A of the Smart Breaker is connected to the DC mains and connection B is connected to the DC sector. The DC sector is discharged at the beginning and the loads are not in operation. During the charging process, all capacities of the components are charged simultaneously.

The Smart Breaker waits for a voltage rise of the DC grid ($U_A$). If this exceeds $U_1$, the precharge process is started after $T_1$ with $S_V = 1$. If the voltage difference $U_A - U_B \leq \Delta U$ undercuts within $T_2$, the switch $S_E$ is closed after the time $T_2$ has elapsed. If the voltage does not undercut $\Delta U$ or if an overtemperature of RV occurs, the precharging process is aborted ($S_V = 0$) and an error is indicated. If precharging is successful, switch $S_V$ is opened after time $T_3$. The precharge sequence is shown in Figure 5.23a. Usually, several DC sectors are connected to the DC grid. When these Smart Breakers were switched on simultaneously in the model systems of DC-INDUSTRIE, the auxiliary power supply was overloaded due to the rather high auxiliary power demand of the development samples (specifically, the pickup current of the contactors for precharging). This can be avoided by offsetting the time when the parallel DC sectors are switched-on. For this purpose, the waiting time $T_1$ is not selected to be the same for all feeders of a level, rather it is increased by between one and ten seconds.

If the DC sector is lightly loaded because certain loads are already drawing power during the precharging process, a steady-state deviation $\Delta U_B$ occurs as a voltage drop at the precharging resistor $R_V$, which the breaker $S_E$ has to be able to bridge, see Figure 5.23b. This load has to be limited in value to limit the equalizing currents. Larger loads may only take up power after the waiting time $T_1$ and after $U_1$ has been exceeded.

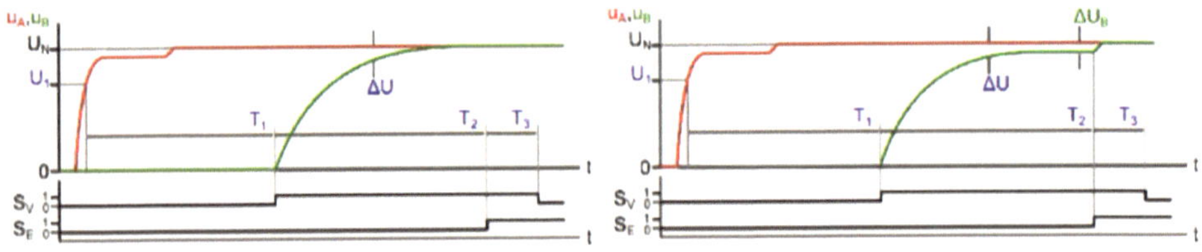


a) Pre-charging sequence of an unloaded DC sector

b) Pre-charging sequence of a lightly loaded DC sector

**Figure 5.23** DC sector precharge sequences

The power for precharging the DC sectors is limited by the design of RV. Each DC sector should be precharged with a maximum factor of $k_V$ = 1/10 of its rated current. This limits the maximum power during simultaneous precharging. Even with significantly more installed power in the DC sectors compared to the feeders, no overload occurs yet.

**Table 5.3** Switching Thresholds and Resulting States of the Switches for Precharging and Operation (therein x = any and * = unchanged)

| State ID \| Name | Voltage $U_A$ | Voltage difference $\|U_A - U_B\|$ | Time $t$ | Power $\|I\|$ | Temperature $T_{RV}$ | Switch $S_V$ \| $S_e$ |
|---|---|---|---|---|---|---|
| 0 \| no voltage (not ready) | $\leq U_{min}$ | x | x | x | x | 0 \| 0 |
| 1 \| voltage available, waiting for precharge | $\geq U_1$ | x | $t \leq T_1$ | $\leq I_{Vmax}$ | $\leq T_{Max}$ | 0 \| 0 |
| 2 \| precharge | $\geq U_1$ | x | $t \leq T_2$ | $\leq I_{Vmax}$ | $\leq T_{Max}$ | 1 \| 0 |
| 3 \| main charge | $\geq U_1$ | $\leq \Delta U$ | $t \leq T_3$ | $\leq I_{Vmax}$ | $\leq T_{Max}$ | 1 \| 0 |
| 4 \| regular operation | $\geq U_{min}$ | $\leq \Delta U$ | $t \geq T_3$ | $\leq I_{max}$ | $\leq T_{Max}$ | 0 \| 1 |
| 5 \| fault precharge, voltage fault | $\geq U_{min}$ | $\geq \Delta U$ | $t \leq T_3$ | $\leq I_{Vmax}$ | $\leq T_{Max}$ | 0 \| 0 |
| 6 \| fault precharge, power fault | $\geq U_{min}$ | $\leq \Delta U$ | $t \leq T_2$ | $\geq I_{Vmax}$ | $\leq T_{Max}$ | 0 \| 0 |
| 7 \| fault undervoltage | $\leq U_{min}$ | x | $t > 0$ | x | x | 0 \| 0 |
| 8 \| fault overvoltage | $\geq U_{max}$ | x | x | x | x | 0 \| 0 |
| 9 \| fault overcurrent | x | x | x | $\geq I_{max}$ | x | 0 \| 0 |
| 10 \| fault over-temperature | x | x | x | x | $\geq T_{Max}$ | 0 \| * |

Furthermore, what has to be stored in the loads of the DC sector is, that they may only load the grid after reaching $U_1$ with a delay of $T_1$. Should the DC sector be disconnected via $S_T$ and then reconnected to the already charged grid, then this can be done without change with the sequence described above. Since the Smart Breaker observes the input voltage $U_A$ only after the switch $S_T$, the precharge time $T_1$ would start at this point directly after closing $S_T$.

The same procedure can be applied without change to all downstream DC sectors at lower levels. The precharge sequences continue with a time delay in all levels.

Table 5.3 lists the states and switching thresholds for precharging and operation. In it, in addition to the preceding functionality, the voltage is monitored; the limits $U_{max}$ = 800 V and $U_{min}$ = 400 V apply. The current is also monitored; it must not exceed $I_{Vmax}$ during precharging, otherwise the charging process is aborted. During normal operation it must not exceed $I_{max}$. During precharging, the temperature at $R_V$ is monitored and if the maximum temperature $T_{max}$ is exceeded, the charging process is aborted. The time t is reset in each case during the transition to a new state.

### Connecting a New Device to a Running DC Grid

If a device has to be connected to a running DC grid, care must be taken that no impermissibly high charging current flows into the device, because the neighboring devices of the same DC sector have no current limitation, however, they do have large DC link capacitors. Possible malfunctions of high charging currents are:

1. The switching device or the contacts of the switch are damaged by high currents.
2. The voltage in the DC sector drops below DC 400 V, so that the devices in the DC sector switch off with undervoltage.
3. The oscillating circuit from the capacitors of the neighboring devices and the wiring in the DC sector charges the DC link of the connecting device to an impermissibly high value, which leads either to overvoltage disconnection or destruction.
4. The Smart Breaker trips.

To avoid this, two basic principles exist:

1. The device or the associated switch ensures that no impermissibly high charging current flows.
2. The corresponding DC sector must first be enabled via the associated Smart Breaker. In the voltage-free state, the device can be connected and then the DC sector can be switched on and precharged via the Smart Breaker.

These principles also apply to the use of maintenance switches. The following procedures are given as examples of technical solution of the specifications:

1. The capacity of the device is so small that no significant charging current flows.
2. The maintenance switch includes a precharge function.

3. The maintenance switch is only a signal switch that activates the switch of the Smart Breaker of the associated DC sector.

Other methods are permitted, provided they comply with one of the two principles mentioned.

#### Recharging in the Event of a Short Circuit or Voltage Dip

In the event of a short circuit, the voltage in the DC sectors drops due to the partial discharge of the accumulators into the short-circuit location. After the faulty DC sector is switched off, the capacities of the "healthy" DC sectors are recharged. A similar situation can occur when there is a high-power demand in a DC sector. Simulations and experiments in the model plants have shown that this recharging can cause the Smart Breaker to trip and shut down.

To ensure high selectivity, various solutions are possible. The following are examples only:

- Grid management would allow to make use of the information about a short-circuit in the parallel DC sector by proceeding for a certain time analogous to the first precharge. However, this would of course limit the power in the parallel DC sectors.
- Another possibility would be to temporarily allow higher overcurrents for the Smart Breakers (if the switches and the connected devices allow this). As a criterion for the load during recharging that the Smart Breaker must allow, the energy requirement of the capacitors during recharging from the minimum transient undervoltage to nominal voltage can be used.

#### Discharge

The discharge of the grid is decentralized in each component that has capacity. For the discharge of the entire grid, the Smart Breakers are separated. By way of appropriating loads (switching power supplies, balancing resistors, ...) or other circuitry each component ensures that the maximum voltage in the device has dropped below DC 60 V within five minutes after shutdown. This discharge must always be active, since discharge must be ensured as soon as neither supply nor storage devices are providing any further energy.

The DC grid itself discharges relatively quickly after all DC sectors have been disconnected because there are no more capacitances on it. For reasons of protection against contact during maintenance work, this discharge to below DC 60 V should take place within 30 seconds. This can only be achieved electrically in a safe way if all DC sectors are disconnected via isolating switches.

If a faster discharge of the grid is required, it is optionally possible to discharge individual DC sectors or the DC grid via a switchable resistor. The precharging resistor in the Smart Breaker may also be used for this, however, only with suitable wiring.

# ■ 5.7 EMC Concept

One of the basic requirements of DC-INDUSTRIE is that already available technologies, standards and operating experience can be used or need to be adapted only slightly. Furthermore, the DC grid is intended to supply industrial plants. In addition, the openness of the system should be guaranteed. These requirements are ensured by the EMC concept, the rules of which are explained in the following sections.

Figure 5.24 shows where EMC interfaces for which rules have been defined may ensue.

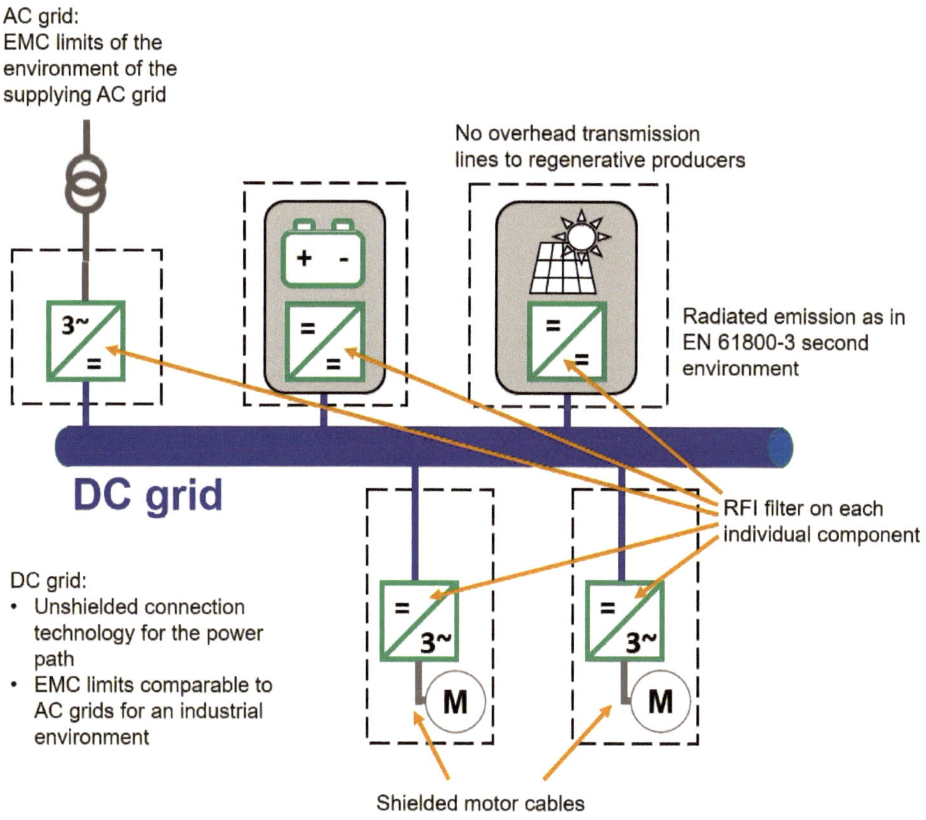


**Figure 5.24** EMC interfaces of the DC grid

First of all, the DC grid is connected to the AC supply grid; here, the EMC requirements demanded by the local grid must be met. Furthermore, some important properties are already included in Figure 5.24, for example the unshielded lines of the DC grid and restrictions of the environment. Furthermore, it is indicated where existing standards and experience are used, e.g., for the installation of converters and motors and for radiated emissions.

Rules must be defined for the DC grid itself. Here, no standard can be used that adequately covers the EMC requirements of industrial environments. For this reason, the interference radiation of test DC grids was examined as part of the DC-INDUSTRIE project. Based on this, EMC rules have been specified, and limit curves defined in order to be able to develop and test components for the DC grid, whereby limit values and measurement methods of AC low-voltage grids, other DC grids and nearby product standards have been adapted in this process.

The following sections explain the EMC rules of DC-INDUSTRIE in detail.

### EMC Interface to the AC Grid

The DC grid is connected to the AC grid via AC/DC converters (supply units). The environment encountered there determines the EMC requirements for conducted interference emission in the range from 0 to 30 MHz. New EMC rules are not required, as there are harmonized standards that can be referred to. Due to the application scenarios of DC-INDUSTRIE, in which most of the energy is converted in electrical drives, it has been determined to apply IEC 61800-3 [DINE2019] for the supply devices. This means that limit values are available with regard to interference emissions and interference immunity to the AC grid.

IEC 61800-3 assigns drive systems to different environments according to their place of use. The first environment, with residential and commercial areas where small businesses and companies are located, is connected to the public low-voltage grid directly. The second environment, in contrast to the first one, is characterized by the fact that each consumer is usually connected directly to the medium- or high-voltage grid via a transformer. Furthermore, the four equipment categories C1-C4 (C1 and C2 for the first environment and C3 and C4 for the second environment) are differentiated and meant to determine rules and limit values respectively.

Due to the different limit values of the device categories, the power converters to be used as well as the filter technology can therefore differ depending on the environment.

### Interference Emissions from DC Lines

The DC grid is implemented with unshielded lines due to the openness of the system and previous experience in AC grids with regard to easy handling. Here, the DC grid differs from proprietary systems, which often prescribe shielded lines. The analyses within the DC-INDUSTRIE project prove that the interference radiation caused by differential mode currents (DMC) can be neglected if the DC+ and DC− lines are laid parallel to each other with a small distance between them. The EMC concept thus requires at least that the DC lines be laid close together. Twisting of the lines, as far as this is constructively possible, is also beneficial.

The interference radiation caused by common mode currents (CMC) is limited to permissible values by EMC capacitors or by the grounding concept of DC-INDUSTRIE. Figure 5.30 shows an exemplary filter arrangement.

### Radiated Emitted Interference

Each component in the system must limit its radiated interference emission. The conditions specified by the environment or corresponding product standards apply here. For electric drives, the limit values of IEC 61800-3 for radiated emitted interference from 30 MHz to 1 GHz, second environment (industry), are used. Measurement methods and limit values do not differ from AC.

### Conducted Interference Emission on a DC Grid

The EMC concept specifies limits for conducted interference emission on DC grid lines as well as their measurement environment. The limit values are designed for industrial environments. They are based on the limit values of and the procedure for low-voltage AC grid. No standards for low-voltage DC grids are yet available for the industrial sector. The limit values presented below therefore apply with reservation, since they were derived from various guidelines. The limit values given refer in each case to the quasi-peak values of the unbalanced interference voltage measured from DC+ to PE or from DC- to PE. The limit values from Figure 5.25 (in V) and Figure 5.26 (in dBµV) must be observed:

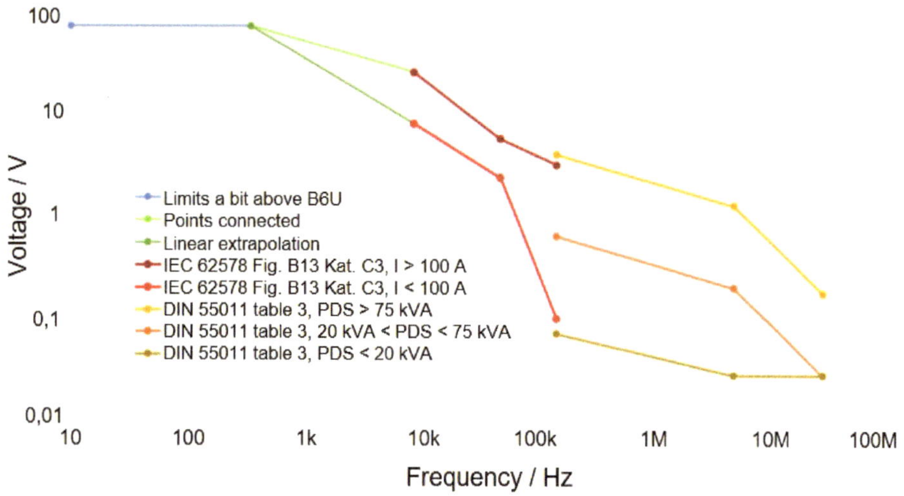


**Figure 5.25** Limits for unbalanced interference voltage for the frequency range 10 Hz to 30 MHz (voltages are shown in VRMS (Quasipeak))

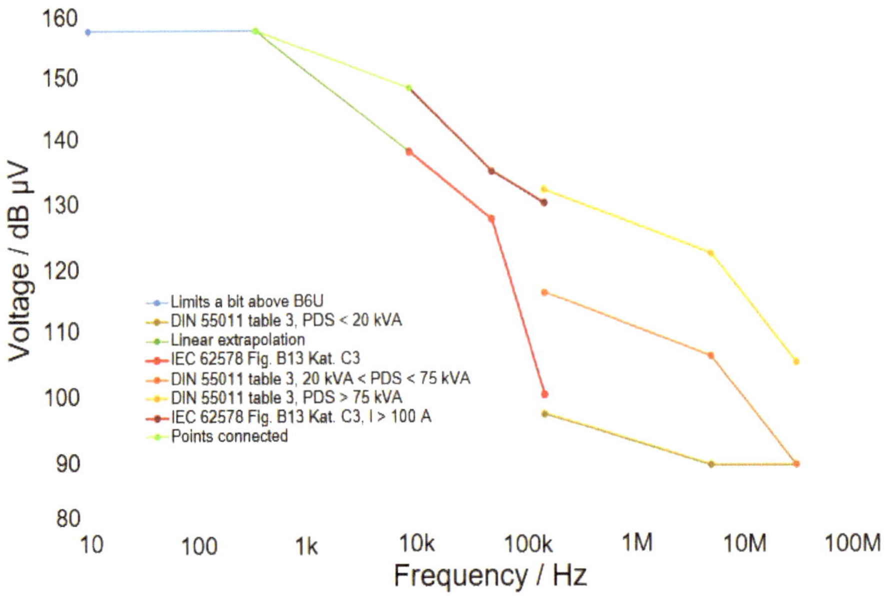


**Figure 5.26** Limits for unbalanced interference voltage for the frequency range 10 Hz to 30 MHz (voltages are shown in dBµV (Quasipeak))

In accordance with current efforts in EMC standardization to close remaining gaps in the frequency range from 2 kHz to 150 kHz, the entire frequency range of conducted interference emission from 10 Hz to 30 MHz is already considered. In detail, the following limit characteristics were defined:

1. 10 Hz-300 Hz: Limit characteristic for low harmonics of the AC grid, caused by the common-mode voltage of uncontrolled 6-pulse rectifiers (B6U). The limit is about 20% above the highest harmonic of an ideal B6U with resistive load.
2. 300 Hz-9 KHz: Here, there is no reasonably transferable range of AC grids, so an extrapolation was carried out using the corner values of ranges 1 and 3.
3. 9 Hz-150 kHz: Here, the limit value proposals of IEC TS 62578 from Figure B13 category C3 for devies with a rated current of less than 100 A and greater than 100 A were adopted, therefore two characteristic curves result.
4. 150 kHz-30 MHz: EN 55011:2017-03 Table 3 [DINE2017] specifies limit values for low-voltage DC grids. These limits are primarily intended for DC lines used in solar power generation applications. In the absence of other limits, they were used in DC-INDUSTRIE and considered plausible. Power ranges for equipment with rated apparent power less than 20 kVA, 20–75 kVA, and greater than 75 kVA are distinguished by their own limit characteristics.

In Figure 5.25 and Figure 5.26 dots indicate the respective corner values of the ranges. The EMC measurement technique for reproducible measurement results is described in the following section.

#### Measurement Results of a Conducted Interference Emission

Within the DC-INDUSTRIE project, two different measurement environments were applied to determine the interference emission of single devices and within a DC grid.

For single devices, a measurement environment consisting of a controlled DC source, a downstream DC grid replica according to CISPR11: 2017-03 Annex I.4.1N 55011 and a radio disturbance voltage test receiver has been used. On the one hand, this environment provides reproducible results. On the other hand, it is possible to fall back on an available grid replica that is already used for DC devices from the PV sector. Figure 5.27 shows an example of the grid simulation from the manufacturer *Schwarzbeck*. With this environment, measurements can be made from a frequency of 9 kHz to 30 MHz.

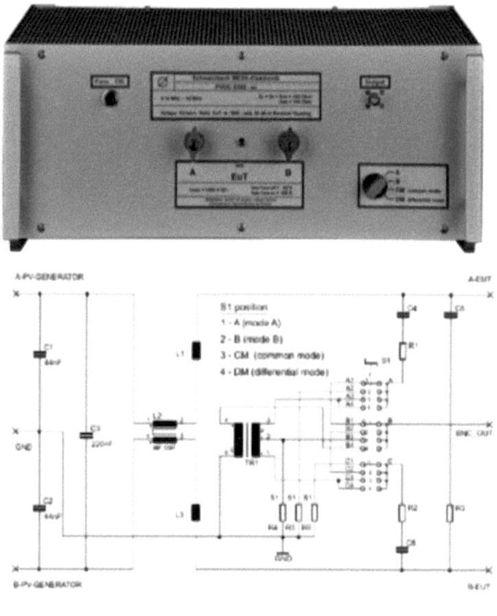


**Figure 5.27**
Grid simulation for DC grids with a circuit diagram (source: Schwarzbeck)

An exemplary measurement of the interference emission of a power converter used in the project is shown in Figure 5.28:

Shown in Figure 5.28 is the interference voltage at the DC+ (DCP) measurement output in the frequency range 150 kHz to 30 MHz. The inverter measured here was a drive inverter. In order to ensure that the voltage undercuts the limit curve (dashed line), minor adjustments were made to the RFI filters.

For measuring the interference emissions in a DC grid or for devices with a rated current above the current carrying capacity of the grid simulation, an EMC RF probe was used together with a radio interference voltage test receiver. Three model systems and some devices in the range from 9 kHz to 30 MHz have been measured with this measurement environment. Figure 5.29 shows a representative measurement in the KHS model system. The limit values according to Figure 5.26 for rated power less than 20 kVA are entered. Devices from many manufacturers were used in this model plant. At the time of measurement, only a few devices were interference suppressed according to DC-INDUSTRIE specifications. It is all the more remarkable that only at one point at 520 kHz a slight exceeding of the limit value can be detected.

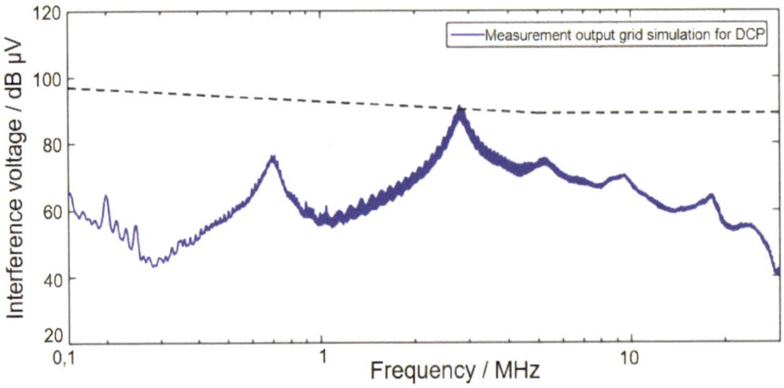


**Figure 5.28** Measurement on grid replica

From the sum of all EMC measurement results, it can be concluded that both the limit value characteristics are suitable, and the effort required for DC interference suppression is justifiable.

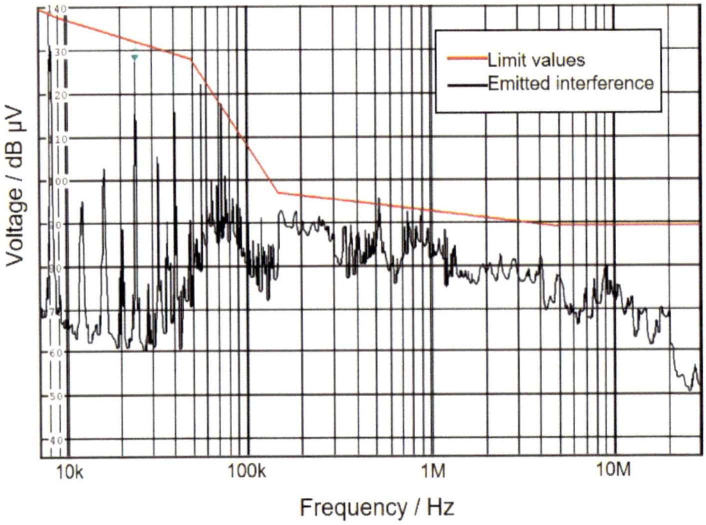


**Figure 5.29** Measurement of conducted interference emission with probe, model plant "Intralogistics System with Handling Stations and Roller Conveyors", DC sector LZ2, measurement during operation of model plant

### DC RFI Filter

In general, equipment that conforms to the proposed limits will require DC RFI filters. Based on the measurement results, it is expected that a single-stage RFI filter consisting of X and Y capacitors and a differential-mode choke can be used. When using differential-mode chokes, the frequency range between 50 Hz and 400 Hz should be avoided for the resonance from the DC choke and DC link capacitance, because excitations from the AC grid due to uncontrolled feeders can be expected in this range.

Since the limit values described there (frequency range 150 kHz to 30 MHz) are established for PV systems, it can be assumed that a DC grid can be designed without shielding if the limit values are observed. Measurements on the model systems during the course of the project have shown that these limits can be complied with taking into account a similar amount of circuitry as would be the case with AC.

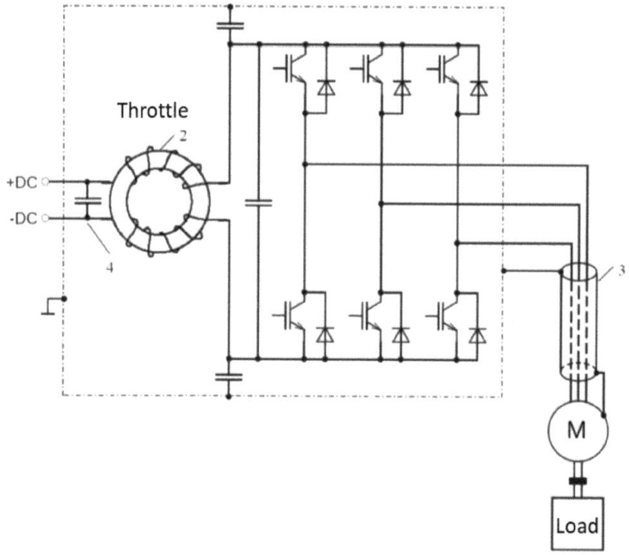


**Figure 5.30** Topology of a drive in the DC grid

**EMC Interface to Motors**: A special feature of a DC grid is that each motor is operated via a DC-AC converter. Thus, there are no motors directly connected to DC, i.e., no so-called mains motors. For the interface to the motor, the established interference suppression measures of converter technology are to be applied. This includes, for example, the shielded motor cables specified by the manufacturers and the manufacturer's specifications for an EMC-compliant design inside and outside control cabinets.

**EMC Interface to Regenerative Generators and Storage Units**: If regenerative generators are integrated, overhead lines are not permitted in this case.

### Interference Immunity

Interference immunity is intended to ensure the robustness of the devices against inter-system interference. With regard to interference immunity, the requirements do not differ from those for AC devices. As far as they are applicable, measurement methods and limit values according to IEC 61800-3 or appropriate product standards are used.

### Leakage Currents

DC leakage currents are negligible due to the symmetrical design. In the event of a first grounding fault, 10 mA is accepted for the DC IT grid.

Pulse-frequency leakage currents do not differ from today's AC solutions. EMC capacitors from DC+ and DC− to PE must be provided on each drive inverter. These must be dimensioned in such a way that the limit values of the conducted interference emissions (frequency range 150 kHz to 30 MHz) are complied with. As a guide value, they must be selected to be a factor of ten greater than the sum of the leakage capacitances at the motor connection.

### Current Harmonics

In the DC grid, low-frequency currents for energy exchange are a desirable feature. However, the current ripple should not reach impermissibly high values that lead to overloading of system components or their parts. From today's point of view, the electrolytic capacitors installed in the drive controllers are the components limiting the current/voltage ripple. In order to ensure trouble-free operation of different interconnected participants in the DC grid in the long term, the dimensioning rule listed in Chapter 5.8 for the DC link capacities is considered sufficient.

## ■ 5.8 Lifetime of DC Link Capacities

An overload of individual DC link capacitors within a DC sector can occur primarily due to strongly differing dimensions of the capacitors per kW connected load. The voltage ripple that the individual capacity has to withstand is relevant here. The following dimensioning rules for the DC link capacities must be observed in the design:

The total capacity of a DC sector results from the sum of the individual capacities of the integrated devices. At the connection of the DC sector, the max. voltage ripple applicable to electrolytic capacitors must not be exceeded. As a rule of thumb, a relative capacitance of at least 25 µF per kW was calculated here in the first approach. However, more detailed considerations have shown that this value only applies to the dimensioning of DC links with electrolytic capacitors, while a much smaller relative capacitance is permissible for film capacitors without affecting the service life.

In the course of the project, therefore, a reference grid was established which simulates the DC link of a grid participant with small DC link capacitance in electrolytic capacitor technology.

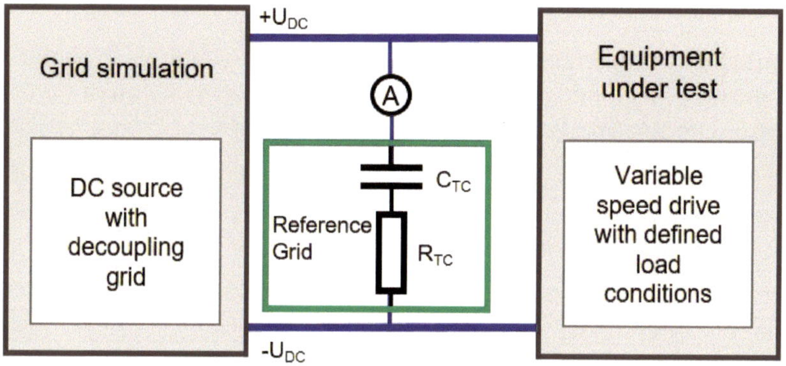


**Figure 5.31** Reference grid for the qualification of DC interconnection participants

This replica (reference grid) consists of a series connection of an "ideal" capacitor ($C_{TC}$) with 1 µF and an "ideal" resistor ($R_{TC}$) with 20 Ω. By means of measuring the current in this reference grid participant, the voltage ripple emitted by the device under test (DUT) can be evaluated (Figure 5.31). In the test setup, a DC voltage source (artificial DC grid) has to be used which decouples the high-frequency currents (e.g., via an inductor) as well as possible.

If the RMS current (effective value) in the reference grid is below 40 mA, it can be assumed that the DUT generates its switching energy essentially from its own DC link. It is therefore suitable for operation in the open DC grid. This emission limit value is derived from the impedance curve of the reference grid. Multiplying the RMS current and impedance gives the frequency-dependent ripple voltage limit curve of typical 400 V electrolytic capacitors (in the frequency range 300 Hz ... 50 kHz). By definition, all frequency components are included in the RMS current. The current measurement must therefore be carried out using a so-called "true RMS measuring device" (bandwidth ≥3 x switching frequency). Alternatively, the voltage curve at the resistor ($R_{TC}$) can be recorded using an oscilloscope and analyzed with regard to the "True RMS measured value".

Instead of a single DUT, the reference grid can also be used to qualify an entire device grid at its connection point to the DC grid. Examples for such a device grid are several parallel coupled decentralized drives, which – due to size and temperature – have a so-called "slim" DC link with film capacitors. Such a decentralized drive would possibly not be DC-INDUSTRIE compliant as a single device, because the built-in film capacitor tolerates and emits a higher ripple current.

The reference grid can also be used in realized DC grid to evaluate the power quality (differential-mode interference voltage between $+U_{DC}$ and $-U_{DC}$) in different DC sectors. In this case, an RMS current of 28 mA must not be exceeded in the reference grid due to the additional capacitances in the feeding DC sectors (immunity limit).

For the calculation of the total rated power of a DC sector, the simultaneity factors of the devices must be taken into account. The total rated power is required for dimensioning the circuit breaker.

The total rated power is also relevant if the actual installed relative capacitance (µF per kW) is to be recorded. Due to the simultaneity factor, the capacitance of a grid participant with a low duty cycle is weighted higher.

To ensure correct project planning of the charging current limitation for the DC sectors, the manufacturers must provide the device-specific value of the DC link capacitance. The simultaneity factor is irrelevant for the charging process.

## Literature

[DINE2019] DIN EN IEC 61800-3-2019-04, Drehzahlveränderbare elektrische Antriebssysteme – Teil 3: EMV-Anforderungen einschließlich spezieller Prüfverfahren (IEC 61800-3:2017)

[IECE2019] IEC EN 61000-2-2, Elektromagnetische Verträglichkeit (EMV): Umgebungsbedingungen – Verträglichkeitspegel für niederfrequente leitungsgeführte Störgrößen und Signalübertragung in öffentlichen Niederspannungsnetzen

[DINE2017] DIN EN 55011:2017-03; VDE 0875-11:2017-03, VDE 0875-11:2017-03, Industrielle, wissenschaftliche und medizinische Geräte – Funkstörungen – Grenzwerte und Messverfahren (CISPR 11:2015, modifiziert); German version EN 55011:2016

# 6 Grid Management

Leopold Ott
Darian Schaab
Alexander Männel
Werner Hinderer

Grid management is a broad term with no generally accepted definition. One possible definition comes from the usage of the term by the energy suppliers, the distribution system operators, and from experience in operating and maintaining AC grid structures within production facilities. Figure 6.1 shows the scope of tasks provided by grid management in the electric utility grid.

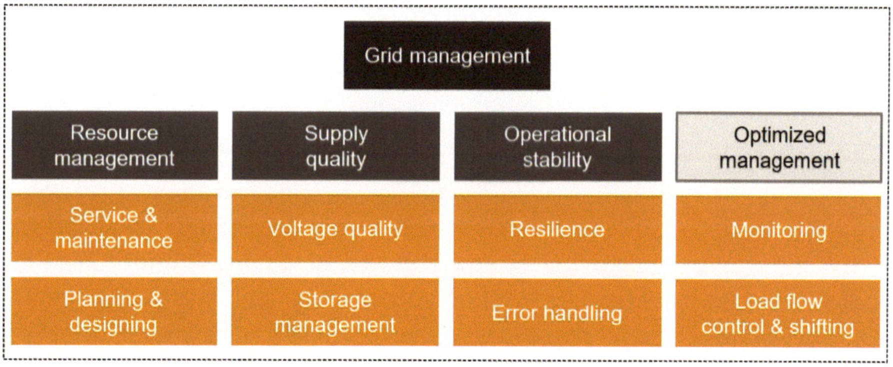


**Figure 6.1** Tasks of grid management in industrial production plants

The grid management tasks shown in gray are necessary prerequisites for reliable grid operation. They have so far been strictly divided between the energy supply company and the end consumer in a hierarchical way. Such a separation means that the industrial end user only has to perform some of the grid management tasks itself, e.g., implement measures to comply with the normative requirements regarding voltage quality. Only when the factory is viewed more broadly as a prosumer with its own energy generation and, prospectively, the option of offering an operating reserve on the market, the role of the end consumer with regard to grid management changes because a large part of the tasks becomes the responsibility of that end consumer.

Tasks that deal with planning, expanding, maintaining and servicing the electrical supply grid are grouped together as operating resources. In industry, the grid as operating resource is projected during factory planning. Since the concept of DC-INDUSTRIE focuses on a transformable supply grid, it must be possible to expand the infrastructure by adding additional storage and generation facilities, but also by adding further consumers.

Everything pertaining to providing electrical energy supply in the quality required by the consumer is summarized under the term supply quality. This includes compliance with limit values for voltage quality, whereby here in particular a simplification can be expected in the future owing to the transition from an AC to a DC grid. The term storage management refers to the implementation of an intelligent charging/discharging strategy for the energy storage units of a grid. The objective might be – for example – to pursue the goals of maximizing consumption of locally generated regenerative energy or of minimizing load peaks for higher-level grid levels.

The term "defined interface" refers to the connections between devices and the supply grid. The grid management describes technical requirements for the connection to the grid and ensures that the integrated devices operate within a defined framework. This framework includes – for example – the definition of voltage limits, communication channels required or optional for operation, and EMC requirements to be met. If the grid is outside this specified framework, disturbances will occur.

All planning and technical activities that are necessary for the safe and stable operation of the grid are assigned to the category "grid stability". This includes – for example – **load flow control**, the task of which is to maintain grid operation at runtime. In the event of *force majeure*, grid management provides fallback scenarios that guarantee a safe state. These measures, usually defined in fault handling, concern – for example – overcurrent protection or measures in the event of lightning.

Grid management can be extended by a targeted control which in turn is based on specific objectives. Such tasks are classified in the category of "optimized operations" management. They include measures for condition monitoring as well as measured value recording, but also mechanisms for monitoring and fault localization in the grid. Another possible goal of operations management is the optimization of energy consumption as a function of production targets and in relation to energy management. Such optimization requires a targeted control of power flows.

In summary, the primary objective of grid management is the targeted distribution and provision of electrical energy in the necessary quality and quantity while constantly observing safety. The secondary objective of grid management is to provide a platform that allows optimized operations management. On the one hand, grid management should enable historically developed generator and consumer structures, while on the other hand it should offer the interfaces for optimized operations management in the future.

## ■ 6.1 Grid Management Participants

Table 6.1 shows the various participant classes. In general, they are clear cut in terms of their main and secondary functions regarding grid management. However, they do show overlaps in terms of energy provision and load flow. Therefore, it is important to clearly differentiate devices from participants in the definition and in the following sections.

**Table 6.1** Presentation of the Main Functions, Secondary Functions as well as the Energetic Classification of the Participants

| Participant class | Main function | Secondary function | Energy supply | Load flow |
|---|---|---|---|---|
| Producer | Supply of energy | Optimized operational management | In terms of time: unlimited or limited energy input. Examples: Interface to AC grid, PV system | Uni- or bidirectional |
| Prosumer | Optimized operational management | Value-added through grid-serving use, e.g., increase in reliability, buffering of braking energy | In terms of time: unlimited or limited energy input. Examples: Interface to AC grid, battery storage. | Bidirectional |
| Consumer | Provision of final energy | Optimized operational management, e.g., through load shifting | In terms of time: unlimited energy consumption | Uni- or bidirectional |
| Smart DC breaker | Reliability through fault management: detection and elimination/ isolation of fault conditions<br>Returning the grid to a safe state | Disconnecting DC sectors and subgrids for maintenance purposes<br>Precharging of a grid section<br>Monitoring of the load flow | Only as a limitation for the energy feed-in during the precharge phase | Bidirectional |

**Producers** are all participants that – within the DC grid – make energy available from an external source as electrical energy, either for an **unlimited** period of time or for a **limited** period of time. On the one hand, such energy could either be provided by a conventional AC supply grid or generated locally by a renewable energy source (e.g., a PV system). Non-regenerative sources are **unidirectional**, regenerative ones are referred to as **bidirectional sources**. Grid management can use the sources for optimized operation management by prioritizing them depending on external influencing factors.

**Consumers** are participants that convert electrical energy for the purpose of using it in a production process. Consumers are part of the value chain and essential for the operation of a production plant. In principle, consumers can draw energy from the DC grid for an **unlimited** period of time. If the consumers are capable of regenerative power supply, they are also referred to as **bidirectional** because they can release part of the absorbed energy again. This energy output is usually significantly smaller than the amount of energy initially taken. If the state of a consumer can be influenced directly or indirectly by grid management, it can be used for optimized operation, e.g., by load shifting.

**Prosumers** draw only a **limited** amount of energy from the DC grid. With a time-delay and minus internal losses, they can return the energy taken almost completely to the DC grid. Prosumers are not essential for maintaining the production process and are primarily used for optimized operational management. The input and output of electrical energy for these participants is controlled by the grid management system. In principle, they are **bidirectional.**

The **Smart Breaker** has the ability to selectively interrupt the current flow between a DC sector/subgrid and the DC distribution grid. On the one hand, this can happen if a fault condition, e.g., a short circuit, is detected on one side of the Smart Breaker. On the other hand, disconnection for maintenance purposes is possible. Further information about the possibility of realizing a Smart Breaker can be found in Chapter 5.5. Furthermore, a Smart Breaker offers the possibility of monitoring the load flow by recording the relevant respective data continuously and by communicating with the DC grid control system.

## ■ 6.2 Options for Load Flow Control

Based on the system concept described in Chapter 5, the following enlarges on the expansion stages of grid management. For all expansion stages, the device classes sources, storage and consumers (e.g., generators, motors and batteries) can be integrated as participants in the grid management with respect to their power electronic converters. A decisive factor in this context is the structure of the communication channels between the devices, which in principle make it possible to distinguish between four different expansion stages:

- No communication
- Decentralized control with control characteristics without additional communication
- Decentralized control with control characteristics and additional non-real-time capable communication option for optimizing operation control
- Centralized load flow control with real-time capable communication interfaces for the distribution of setpoints

It should be noted that devices with a communication interface that allows their integration into both a centralized and a decentralized load flow control concept are expressly permitted because of their flexibility to meet the requirements of the respective user.

### Decentralized Grid Management *without* Control Characteristics and without Communication ("Basic Grid")

This form of load flow provision is completely decentralized, i.e., there is no overall coordination of the feed-in power of the individual sources during the operation of the system. The generators provide an unregulated or regulated DC grid voltage or a defined feed-in current level, e.g., according to the current maximum feed-in power of a DC/DC converter for a PV system. All other devices follow their own operating strategy according to their respective intended purpose.

Such an interconnection represents the simplest operating mode. It is referred to as the **basic interconnection**. The power flow is not controlled and can therefore only be influenced by prior parameterization of the devices in the switched-off state. However, with this type of grid management, special attention must be paid to ensuring that the grid is not overloaded in the event of feed-in from multiple sources. This can be done – for example – by selectively reducing the feed-in power of a source when certain voltage limits are exceeded.

### Decentralized Grid Management *with* Control Characteristics and *without* External Communication ("Decentralized Group Control")

In contrast to decentralized or isolated control, preconfigured control characteristics are set up in the converters (AIC and DC/DC converters). The supply system in the expansion stage "**decentralized group control**" provides information about the grid status.

Each participant measures and regulates decentrally in relation to the terminal voltage on the DC grid side. The power flow control is realized via so-called control characteristics (cf. Chapter 5, system concept). In addition, the information shared over the voltage band can be used for an optimized operation management advantageously for the control of producers and prosumers. A design of the control characteristics and the configuration of the DC grid participants can realize the advantages of optimized operation management without additional communication.

Basic features:

- Converters with a control characteristic are referred to as active. Converters without control characteristics (passive converters) are considered as faults.
- The control characteristics of all active converters are tuned to each other depending on the existing grid configuration, i.e., the central tuning is performed offline before commissioning the system.
- Sources, storages, consumers without converters are – like passive converters – considered as faults.
- Power flow control is based on the control characteristic built into the active converters. Depending on the voltage applied to the converter, the current flow (and thus the power) is independently controlled in a decentralized manner by the converter concerned. The relationship between current and voltage in the control characteristics is continuous and linear.
- No additional communication between the participants is required for the basic operation of the control.
- The power flow control is not influenced by higher-level systems (SCADA, MES or other higher-level grid management participants).
- Optimized operation management (power flow control) can be achieved by tuning the control curves accordingly during commissioning.

#### Decentralized Grid Management *with* Control Characteristics and *with* Communication ("Extended Group Control")

If an additional communication structure via a field bus to the DC grid participants is available, a DC grid can be implemented in the **extended group control** expansion stage. This expansion stage allows interactions with the process PLC, external control of the grid as well as monitoring of the grid status. The power flow of all devices can be influenced via a central management unit (grid manager) or an agent system in a user-friendly way at runtime by modifying the control characteristics. The adjustment and optimization of the interaction of all devices with control characteristics runs via the central grid manager or decentralized via agents communicating with each other. These agents know the situation in the immediate environment due to additional sensors (e.g., temperature) or the status of their own production process (e.g., material availability, order situation, malfunctions). By including its own environment information and that of neighboring agents, each agent can individually adapt its control characteristic and thus its energy consumption to the production environment. There are links here to cyber-physical systems in the “Industry 4.0” environment.

The system can be optimized during operation by configuring the control parameters. While all active producers and active prosumers are integrated into the power flow control, there is an additional possibility to integrate active consumers into the smoothing of load peaks.

Basic features (such as decentralized group control with additional freedoms):

- Grid management can modify parameters of control characteristics on active converters during runtime through power flow control.
- Grid management can be centralized via a controller, a software manager (PC based), or decentralized via intercommunicating agents to be installed on field devices.
- Work sharing with an existing process PLC and connection to higher-level systems (SCADA/MES) are possible.
- Not all feeding devices have to be integrated into the communication infrastructure, since the characteristic-based control works preconfigured.

**Central Grid Management via a Higher-Level Control System ("Central Grid Control"):**

Central systems are controlled via an open and standardized control structure (e.g., PLC according to IEC 61131-3). The central controller (e.g., programmable logic controller) regulates the DC grid to a specified DC voltage. The voltage can be changed variably during runtime. The central system can be adjusted in its dynamics and its target values in real time via various control parameters. All target values for the DC grid participants are taken over by the central control system. All feeding devices must be integrated into the communication infrastructure for power flow control.

Basic features:

- Central communication with the field devices involved is required. For this purpose, they are connected to a dedicated power flow control (as distinct from the process PLC). As a rule, both controllers share the same communication bus.
- Sources, storage, loads without converters and converters without communication interface are considered as faults.
- Field devices without converters with communication interface (sources, consumers, storage devices, measuring devices) can be included in the power flow control if this is useful for the power flow control.

- The central power flow control makes specifications for the current flow/power depending on the voltage at the individual converters. It ensures a balanced power in the grid, taking into account predefined priorities.
- Control parameters and priorities can be changed at runtime (e.g., for the purpose of feedforward control). The power flow control receives the corresponding parameter specifications from higher-level systems (SCADA, MES).
- All additional state management functions (switching off/on of field devices, error messages, precharging coordination, AC mains failure management) are the task of the process PLC.

## 6.3 Recommendations for the Design of Power Sources in the Context of a Characteristic-Based Control System

For the verification of the small-signal stability in connection with the characteristic-based load flow control methods, the methods of Ott [Ott2015] can be used, where the main focus is on the compliance with the so-called passivity criterion. Here, the DC grid is decomposed into two subsystems consisting of the producer subsystem, which includes all sources, and the consumer subsystem, which includes all loads. Both systems interact via a common node, the DC grid. A schematic representation of this model can be found in Figure 6.2, where it is assumed that the distribution grid contains n producers and m consumers.

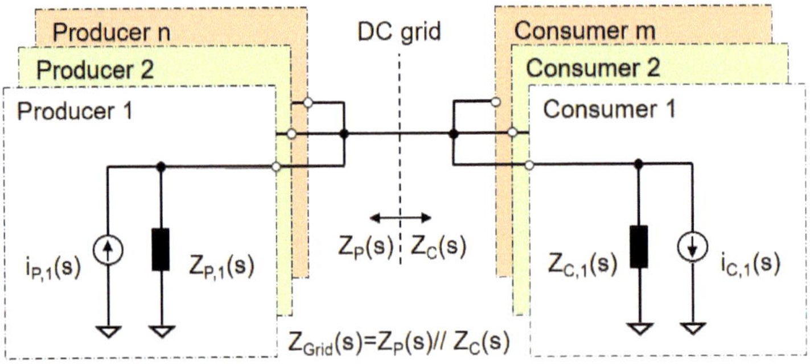


**Figure 6.2** Decomposition of a DC grid at the common node DC bus into subsystems for producer and consumer

The dynamic behavior of each individual producer and consumer is described by a complex impedance $Z_{P,n}$ and $Z_{C,m}$, respectively. These impedances are defined as dependent on an operating point in order to additionally represent the large signal effects, such as the operating point dependence of transfer functions. Both subsystems can then be described by the parallel connection of the individual impedances of their subcomponents, which is expressed by the equivalent impedances $Z_P$ and $Z_C$. The equivalent grid impedance is then again calculated from the parallel connection of $Z_P$ and $Z_C$.

The passivity criterion is a sufficient, but not a necessary criterion to fulfill the small signal stability for a given operating point. The criterion is satisfied if $Z_{Bus}$ satisfies the condition for passivity. On the one hand, this means that $Z_{Bus}$ must not have any pole positions in complex half plane on the right and, on the other hand, the phase response must lie between ±90°. Passivity for $Z_{Bus}$ can be achieved by force by ensuring passivity also for $Z_P$ and $Z_C$. However, there is a restriction for $Z_C$. Since the majority of loads in a DC grid consume constant power on the line side, these devices exhibit active behavior within the bandwidth of their output-side regulation. In relation to the input impedance, this shows ohmic behavior with a negative sign. However, due to the input circuitry of the load, either consisting of a capacitor or a higher order filter stage, the dominance of the active behavior is limited to the range of very low frequencies.

To rule out stability problems in the frequency range in question, there must be no intersection between the magnitudes of $Z_P$ and $Z_C$ in the frequency range in question. Finally, this implies that the magnitude of $Z_P$ for low frequencies must be significantly smaller than that of $Z_C$, which can always be assumed in practice. Outside the active range, the frequency response of $Z_C$ is determined only by the passive input circuitry, which always ensures passivity for this subsystem.

On the other hand, fundamental considerations for ensuring the passivity of the elements of $Z_P$ can be found in [Ott2019], which can also be used as a dimensioning basis for the producers of the DC grids within the DC-INDUSTRIE project. For the active part of the producer, the following equivalent impedance $Z_{D,n}(s)$ is defined:

$$Z_{D,n}(s) = R_{D,n} \cdot \left(1 + \frac{s}{\omega_{TP}}\right) \cdot \left(1 + \frac{s}{\omega_i}\right) \tag{6.1}$$

While the virtual resistance $R_{D,n}$ reflects the slope of the control characteristic of a device at the operating point under consideration, the angular frequency $\omega_{TP}$ defines the position of the corner frequency $f_{TP}$ of a measurement low-pass filter for the DC line-side voltage. The design of the measurement low-pass filter is application-specific and serves to suppress interfering high-frequency components in the line-side terminal voltage, which can be caused, for example, by weakly damped line resonance points. A value of the filter corner frequency between half and one fifth of the bandwidth of the inner current control loop $f_i$ is suitable for many applications. $f_i$ is usually between one sixth and one tenth in relation to the switching frequency. The mapping of $f_i$ in the frequency domain is again done with the help of the angular frequency $\omega_i$.

The passive part of the output impedance $Z_{P,n}$ is determined in particular by the DC line-side capacitor $C_{G,n}$, for whose impedance the following simplified expression is applied:

$$Z_{C,n}(s) = \frac{1}{s \cdot C_{G,n}} \tag{6.2}$$

The impedance $Z_{P,n}$ is finally calculated from the parallel connection of $Z_{D,n}$ and $Z_{C,n}$. Looking at the definition of $Z_{D,n}$, it is initially noticeable that there is no passive equivalent for this impedance, since there are two cascaded zero points whose asymptotic phase value for high frequencies corresponds to 180°. The idea of the design recommendation is now to dimension the capacitor $C_{G,n}$ so large that the influence of the inner current control loop disappears. The capacitor is dimensioned according to the following Equation 6.3:

$$C_{G,n} = \frac{1}{\omega_{il} \cdot R_{D,n} \cdot k_c^2} \tag{6.3}$$

The influence of the inner current control loop on the output impedance disappears the lower the value for the parameter $k_c$ is set. For these low values of $k_c$, the expression of $Z_{P,n}$ then finally simplifies to Equation 6.4, which, as desired, corresponds to the description of a 2nd order passive system.

$$Z_{P,n}(s) \approx R_{D,n} \cdot \frac{\left(1 + \frac{s}{\omega_{TP}}\right)}{1 + s \cdot R_{D,n} \cdot C_{G,n} + s^2 \cdot \frac{R_D \cdot C_{G,n}}{\omega_{TP}}} \tag{6.4}$$

A comparison for the prediction of the impedance curve by the approximated representation according to Equation 6.4 with the complete analytical calculation with consideration of the influence of the inner current control loop is shown for an exemplary converter system in Figure 6.3. A bandwidth of 10 kHz is used for the inner current control loop, a cutoff frequency of 3 kHz for the measurement low-pass filter, and a value of 0.5 Ω for the virtual resistor $R_{D,n}$. The parameter $k_c$ is used as a running parameter.

As can be seen, the agreement between the approximated representation and the exact model is sufficiently good from a value for $k_c$ of 0.5. For larger values, the resonance peak in the exact model increases significantly more than predicted by the approximation. For values of $k_c$ greater than 0.7, an increasing violation of the passivity criterion is seen by phase values below −90°. For $k_c$ equal to 1.5, there is also a double pole in the right half-plane, which is evident from the jump of 180°

in the phase response in the region of the resonant frequency and – therefore – cannot be recommended as a practical design.

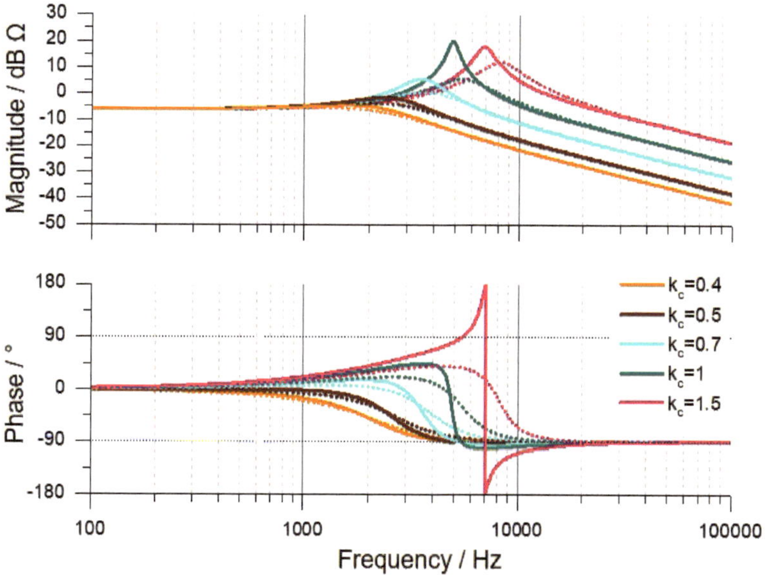


**Figure 6.3** Complex output impedance of a producer for different $k_c$. Comparison of the approximated model (dotted) with the exact representation (solid)

In addition to ensuring the passivity of the output impedance, it is also necessary to evaluate the quality factor in order to estimate the dynamic characteristics when subjected to load jumps. Since, depending on the design, this is a system with an order of 3, which contains zeros as well as poles, the procedure according to Hämmerle [Hämm2015] must be used to determine the quality factor. Based on the measurement of the output impedance, the quality factor $Q_G$ is determined by the group delay $\tau_G$, which is the derivative of the phase response. Analytically then the following applies:

$$Q_G = \tau_G \cdot \pi \cdot f \tag{6.5}$$

For the previously considered converter system, the frequency-dependent curve of the quality factor is shown in Figure 6.4 for selected designs of the DC line-side capacitor.

In general, a value for the quality factor below the value of one for an individual device is desirable in order to keep the magnitude of the overshoot or undershoot to a maximum of 20% of the magnitude of a load boost and to ensure robustness against the sum of the other capacitances and, associated with this, the resonance points, in the DC grid. Even if all other devices in the grid hardly contribute any grid-side capacitance, this ensures acceptable transient behavior.

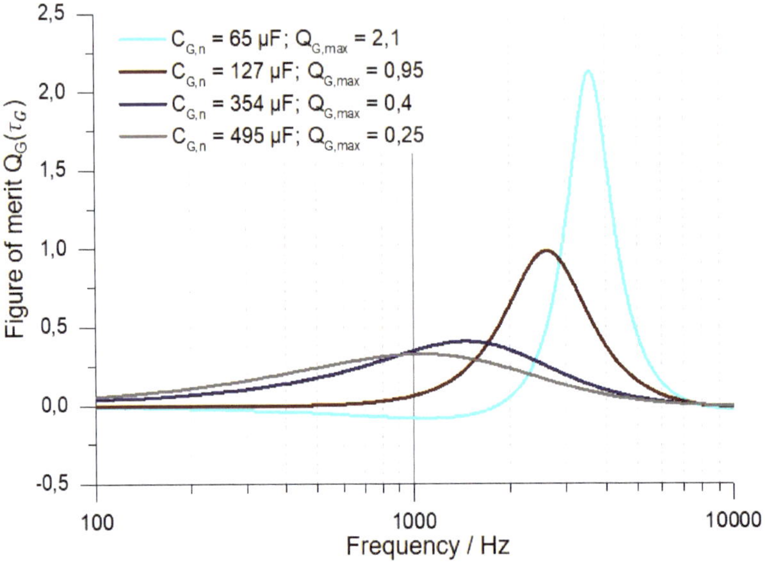


**Figure 6.4** Frequency dependence of the quality factor of a producer of $C_{G,n}$

Often, in real applications, system-internal delays caused by the digital implementation of the control loops and the characteristic control further increase the quality factor. In particular, DC/DC converter systems that fall into discontinuous operation in the partial load range then experience a drastic reduction in the bandwidth of the internal current control loop, which leads to a sharp increase in the quality factor at these operating points and worsens the transient behavior of the system [Ott2019]. Dimensioning the DC grid-side capacitor according to Equation 6.3 creates some robustness against these degrading influences.

In the course of the development process, it is recommended to measure the complex output impedance of a producer under all possible operating conditions in order to characterize its behavior as precisely as possible in the frequency range and thus to obtain conclusions about possible stability problems during operation on the grid. A method that provides reliable results is described by [Ott2015]. The basic schematic circuit diagram of the possible injection of an excitation signal with changing frequency (Chirp signal) can be found in Figure 6.5. The coupling of the excitation signal can be either inductive or capacitive. The frequency response of the complex output impedance is obtained by dividing the frequency-selectively recorded quantities of the voltage $v_{p,n}$ and the current $i_{p,n}$.

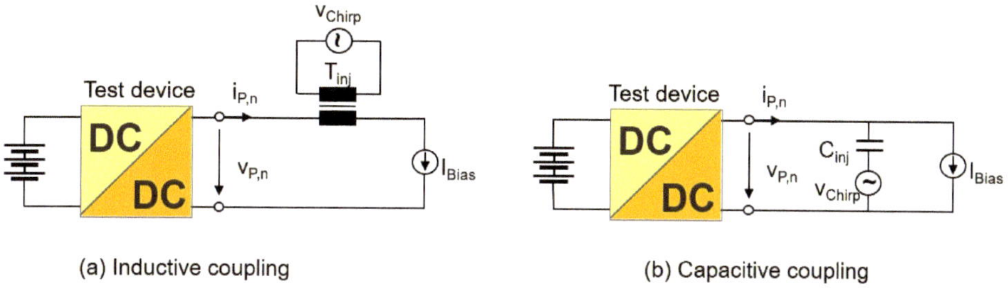


**Figure 6.5** Schematic representation of impedance measurement methods using Chirp signals

The current source with the DC value $I_{Bias}$ is used for setting the operating point. As mentioned before, especially for DC/DC converter systems where the gapped operating mode can occur in partial load operation, the metrological recording of the output impedance at these operating points is important. As an example, two recorded impedance measurement curves for a DC/DC converter are shown in Figure 6.6 for two operating points, one in non-gap (abbrev. CCM) and one in gap (abbrev. DCM) operation. The value of the virtual resistance is 1 Ω. Using a switching frequency of 30 kHz results in practical values for the bandwidths of the inner current control loop and DC line voltage measurement filter of 5 kHz and 1.5 kHz, respectively.

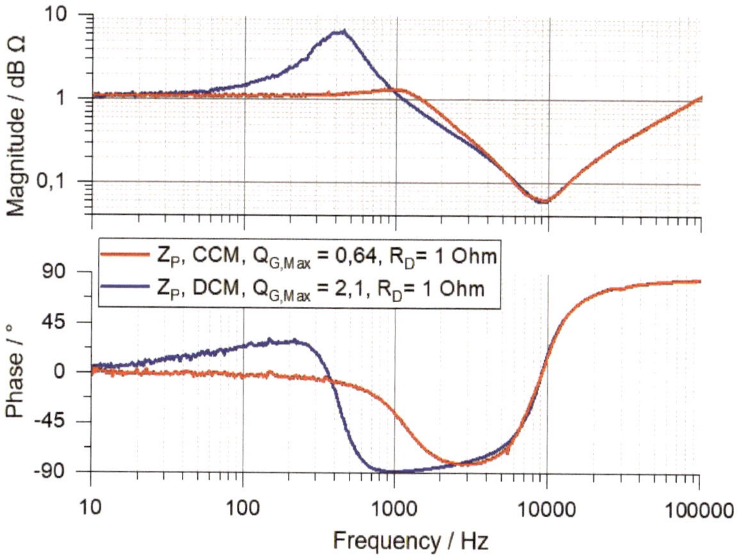


**Figure 6.6** Example of recorded impedance measurement curves for a producer with characteristic-based control at different operating points

The value of the DC line-side capacitor was estimated using Equation 6.3, where a value of 0.5 was applied for the parameter $k_C$. The non-gap operation shows very nicely the compliance with the passivity condition and a desired very low value for the quality factor. For the operating point in gap mode, on the other hand, the expected increase of the quality factor and thus a degradation of the transient behavior is shown. The tolerability of this degradation depends on the other components of the DC grid and must be assessed on a case-by-case basis. It must also be noted that the system behavior shown in Figure 6.6 becomes increasingly inductive for frequencies above 5 kHz, which can be explained by the EMC output filter circuitry.

## 6.4 Implementations of Decentralized Group Regulation in Practice

The decentralized group control was implemented on a CNC processing center from *HOMAG*. It was first examined conceptually using a simulation model that depicts the energy flows of the plant. Based on this, an optimal operating strategy for the battery storage was derived.

In normal operation, the lithium-ion energy storage system is used to absorb braking energy, which, as it stands today, is converted into heat in conventional CNC machining centers by means of braking resistors. The operating strategy of the energy storage unit is to absorb as much energy in the course of a machine cycle as is released again in the course of the cycle. This ensures that the state-of-charge of the battery storage is the same at the beginning of the cycle as it is at the end of the cycle. Since the CNC machining center is operated with variable cycles, a fixed characteristic curve inevitably leads to the lithium-ion energy storage being either overcharged or over-discharged in the course of many machine cycles.

To prevent this, two characteristic curves have been introduced for the operation of the energy storage system: one that results in the battery storage system being charged over the course of many cycles and one that results in the battery storage system being discharged over the course of many cycles. The grid management switches back and forth between these two characteristic curves. The state-of-charge of the battery storage is used as a switching criterion. If the state-of-charge exceeds an upper threshold, e.g., SoC (t) = 80% (state-of-charge), the system switches to the characteristic curve that predominantly discharges the battery storage. When the state-of-charge reaches a lower threshold, e.g., SoC (t) = 30%, the system switches back to the characteristic curve that predominantly charges the battery storage [Männ2019].

The energy storage additionally assumes the function of an uninterruptible power supply in case of an AC mains failure (Figure 6.7). In the event of an AC mains failure, the energy storage system provides the complete power requirement of the machine - in the example a CNC machining center - for a limited number of machining cycles.

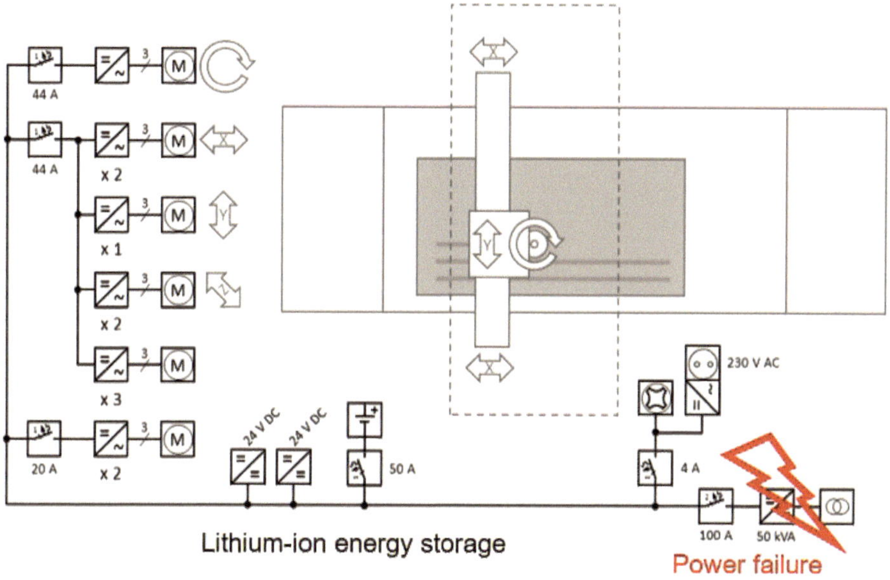


**Figure 6.7** System test "AC mains failure" in a CNC machining center

The method of characteristic curve switching was implemented in a PLC and initially verified using a laboratory demonstrator. Figure 6.8 shows the two characteristic curves, mainly discharging (a) and mainly charging (b). Figure 6.8 (c) shows a test production cycle. The test production cycle was examined in both simulation and experiment using a laboratory demonstrator to test the correct operation of the characteristic switching. In Figure 6.8 (c), the voltage curve of the DC grid shows two different phases with two different fluctuation widths for $u_{DC}$. Initially, the "predominantly discharged" characteristic was set. With this characteristic, the energy storage has a significantly higher energy contribution to the total energy demand of the DC grid than with the "predominantly charging" characteristic. From t =1050 s, the system switches to the "predominantly charge" characteristic. Now the energy storage system hardly contributes any energy to the total demand. The energy storage is only charged during recuperation energies. Since the total demand is now mainly covered by the grid supply, the overall characteristic slope is lower. As a result, the fluctuation range of the voltage from t > 1050 s is significantly higher.

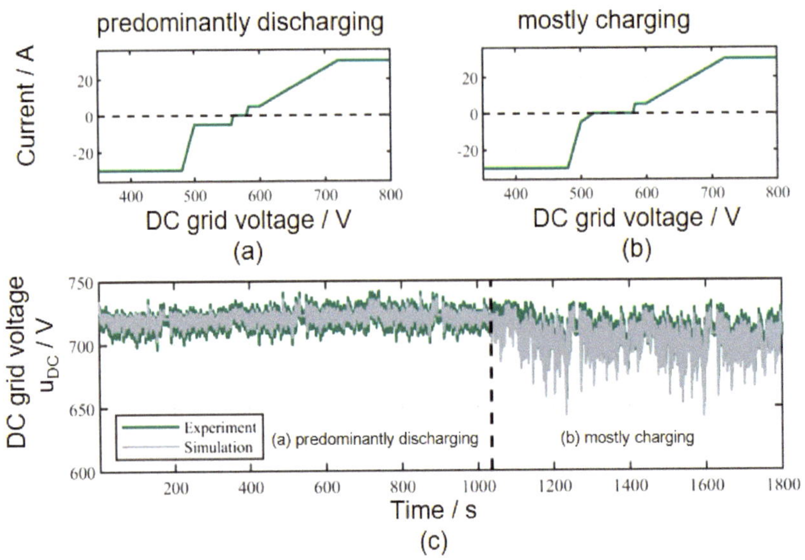


**Figure 6.8** (a) Predominantly discharging characteristic; (b) Predominantly charging characteristic; (c) Voltage curve in a DC grid with characteristic switching

Two different scenarios in a CNC machining center were examined. In scenario 1 "AC mains failure with controlled shutdown", the production cycle is brought to a standstill and the model plant is shut down in a defined manner. Scenario 2 "AC mains failure with mains return", on the other hand, shows that the DC mains returns to normal after the AC mains returns. During the period of the grid failure, the system is fed from the battery storage. To simulate a grid failure, the utility (unregulated rectifier) was disconnected from the grid. Below are measurement curves for the two scenarios.

### Scenario 1 "AC Grid Failure with Controlled Shutdown"

Figure 6.9 shows the voltages and currents of the DC grid and two DC sectors (utility and energy storage) as well as some PLC variables and the state-of-charge (SoC) of the energy storage. At time t = 0 s, the AC grid failure occurs. From this time on, one can see a drop in the voltage in the DC grid. This is due to the characteristic control of the DC grid. In the case of AC mains failure, only the characteristic curve of the DC/DC converter is effective. The utility, on the other hand, no longer provides any power. This can also be seen in the current characteristics of the energy storage device (blue) and the supplier (red). From t = 0, the current of the supplier is zero. The controller has also detected the power failure. Accordingly, the Boolean variable "AC mains failure" switches from TRUE to FALSE at t = 0. The state-of-charge (SoC) of the battery storage is also read by the PLC. This decreases continuously - as expected after the mains failure - since all power for the DC grid must now be provided via the energy storage.

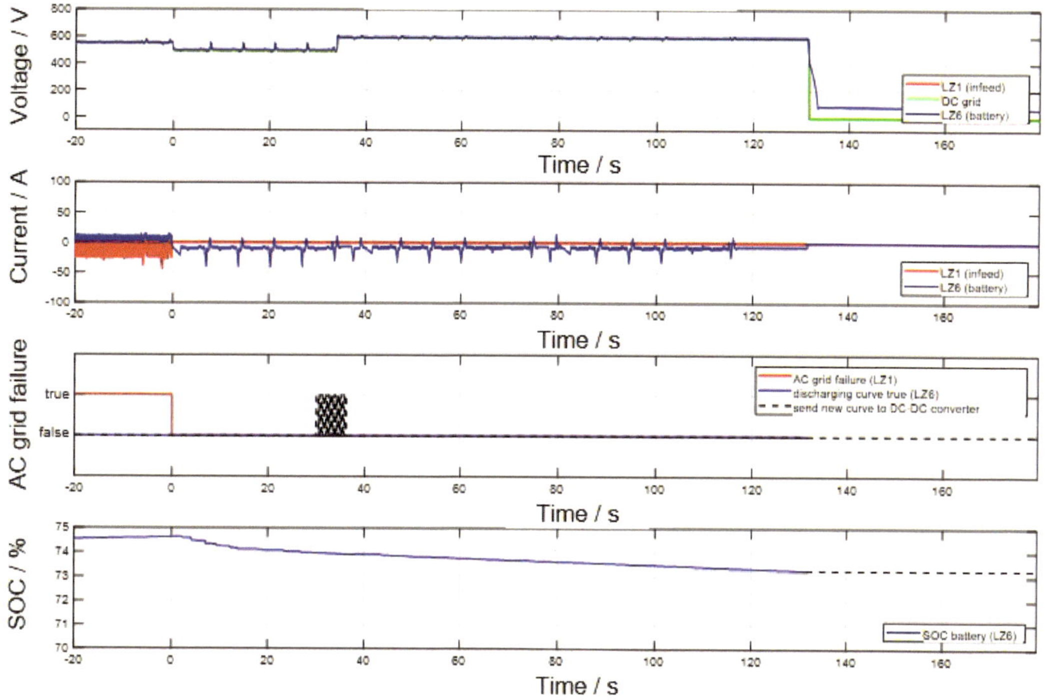


**Figure 6.9** Simulated AC power failure with controlled shutdown in a CNC machining center

In order for the supplier to – in principle – be able to reconnect to the DC mains in the event of an immeadiate AC mains recovery, the DC mains voltage must be above the mains peak of the AC conductor-to-conductor voltage, as otherwise very high equalizing currents would flow during connection, which could lead to false tripping of protective elements. To bring the DC mains voltage back to a suitable level, the PLC switches the characteristic curve of the DC/DC converter to a special characteristic curve 30 s after mains failure (see Figure 6.9: third graph, black dashed line). This special characteristic is activated only in case of power failure and can be seen in Figure 6.10. After the transmission of the new characteristic curve has been successful, there is an increase of the voltage in the DC grid at about 35 s. The voltage is then increased in the DC grid. This procedure always takes place, regardless of whether there is a return of the AC grid or a shutdown of the production line.

From about 120 s, the current and voltage waveforms in Figure 6.9 no longer show any deflections. This indicates that the production cycle has been switched off. At approx. 130 s, the main switch is finally opened. The current flow of the battery comes to a standstill. The voltage in the DC grid drops to zero. In the DC sectors, a residual voltage remains which is only slowly reduced due to the capacities installed there. After this process has been completed, the state-of-charge of the battery naturally remains constant.

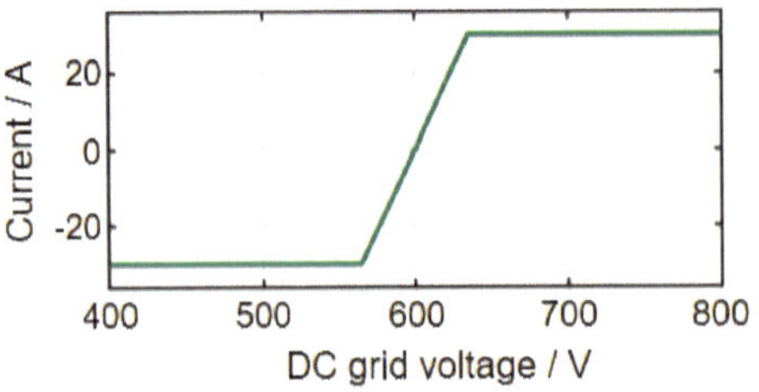


**Figure 6.10** Control characteristic in case of power failure for a CNC machining center

### Scenario 2 "AC Mains Failure with Mains Recovery"

Figure 6.11 shows the curves for scenario 2 "AC mains failure with mains recovery". Up to power recovery at approx. 60 s, the curves are very similar to those of scenario 1. After mains recovery, the variable "AC mains failure" (red line, third graph) is switched to TRUE. Shortly after, the rectifier is switched on again (negative red current peak). However, since the characteristic curve of the DC/DC converter has been shifted significantly after the mains failure and is above the mains peak of 565 V, power is still drawn exclusively from the energy storage device. For this reason, one minute later, i.e., at 120 s, the characteristic curve of the DC/DC converter is shifted back to the original state. After 120 s, the original state of the system is then restored and the state-of-charge of the battery also begins to rise again.

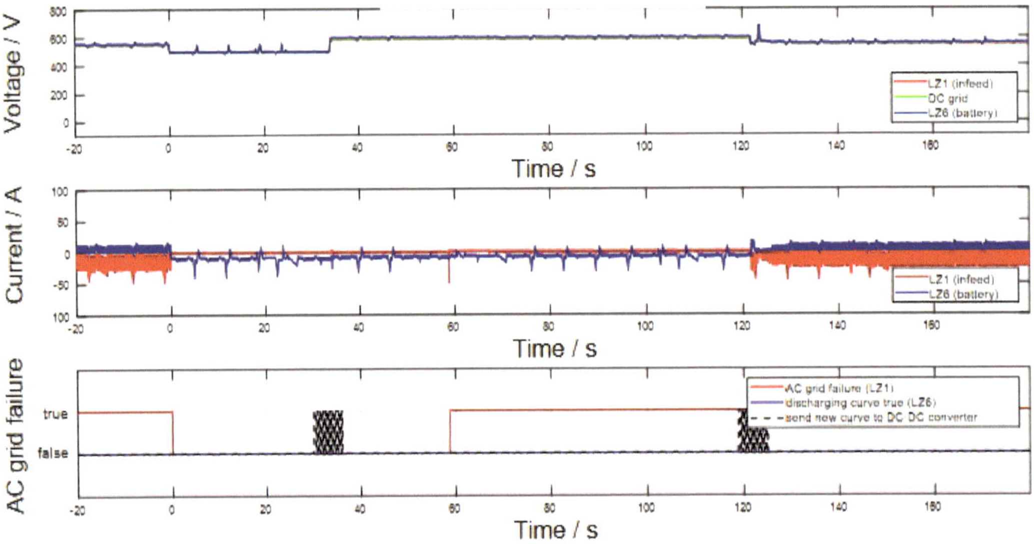


**Figure 6.11** Power failure with power recovery

## 6.5 Charge Control of Electrochemical Energy Storage Devices

When a prosumer is integrated into the DC grid - as shown in Figure 6.12 - the question arises as to how the load $P_{load}$ is shared between the different energy sources. Because energy storage systems have only a limited energy content, they must be recharged when a discharge cycle from the grid is requested in order to be ready for the next phase of energy provision.

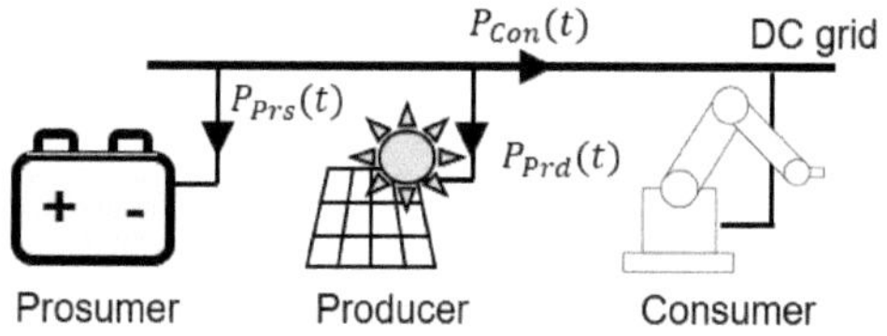


**Figure 6.12**
DC grid with storage

A strategy to operate the storage is shown in Figure 6.13. Based on the load on the grid $P_{Con}$, the power is divided between the prosumer $P_{Prs}$ and the producer $P_{Prd}$. The strategy shown here uses storage to minimize the peak load on the grid. When the load at the utility rises above the limit $P_{Prd,PS}$ (**P**eak **S**having), which determines the maximum peak load from the utility grid, the storage (prosumer) begins to take the additional load. During this discharge phase, the state-of-charge of the prosumer decreases. Figure 6.14 shows the matching design of the control characteristics. The peak power of the grid $P_{Prd,PS}$ is reached at the lower prosumer voltage $U_{Prs,U}$. The nominal voltage of the grid is denoted by $U_n$.

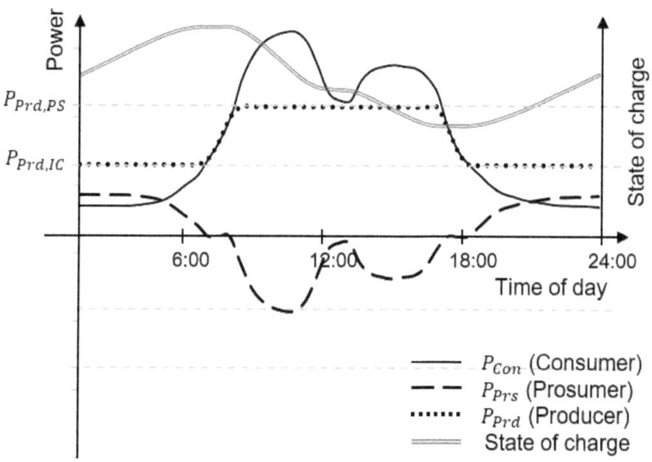


**Figure 6.13** Power curve in the DC grid with storage

As shown, the energy provided by the prosumer must be loaded back into the storage when the load demand is lower. This means the state-of-charge is raised again with energy provided by the producer. Prosumer energy provision can be understood as a measure to decouple energy demand and retrieval by the producer over time. A power limit $P_{Prd,IC}$ (**I**nternal **C**harging) can be defined, which specifies below which load level the storage is recharged. Translating this behavior into a control characteristic results in a plateau starting at the upper producer voltage $U_{Prd,O}$ when the load falls below $P_{Prd,IC}$. Figure 6.13 shows that the state-of-charge increases during these periods and that it returns to the initial state in the optimum case.

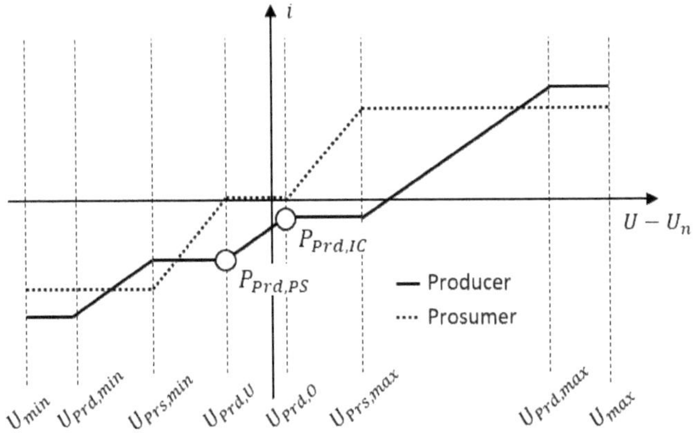


**Figure 6.14** Exemplary design of the characteristic curves for a system with prosumer and producer

However, this state-of-charge control based on the control characteristic programming only works if the forecast of the load curve is known with sufficient accuracy. In case of an unplanned deviation, the design fails, and the prosumer is not or only limitedly available for the next energy supply. To address this problem, the storage system must pursue state-of-charge control as a secondary control objective. Two basic approaches can be distinguished.

The first approach pursues the active shifting of the characteristic curve. The effect on the characteristic curve is shown in Figure 6.15. If the state-of-charge of the prosumer deviates from a specified target range, the characteristic curve is shifted by a constant offset γ. This causes the prosumer to draw additional power from the DC grid in the voltage range of the plateau.

The additional current ensures that the state-of-charge of the storage unit is kept at a certain level. The advantage of this control is that it is robust against stochastic fluctuations in the consumer's power consumption.

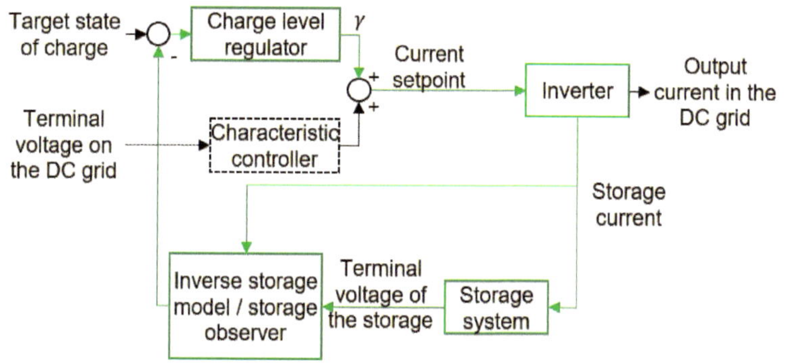


**Figure 6.15** Shifting the storage characteristic with a current offset

Figure 6.16 shows the structural diagram of the prosumer control with characteristic control loop and secondary state-of-charge controller. The state-of-charge control loop is shown in green. The converter provides a current on the storage side, which leads to a change in the terminal voltage of the storage. From the storage current and terminal voltage, the state-of-charge of the storage can be determined via an inverse storage model or an observer. The state-of-charge controller calculates the characteristic shift γ from the control deviation of the state-of-charge.

The characteristic controller and the storage controller act against each other during operation, since a request for energy from the prosumer leads to a drop in the state-of-charge. The state-of-charge controller reacts to this by shifting the characteristic curve so that the requested current setpoint of the overall control loop drops to zero over a longer period of time.

The fact that the storage nevertheless participates in the grid control is due to the fact that the characteristic control loop and the state-of-charge controller are dominant in different frequency ranges. The aim is to design the state-of-charge controller to be dominant in the low-frequency range so that long-term charging cycles that exceed the maximum permissible energy content of the prosumer are avoided.

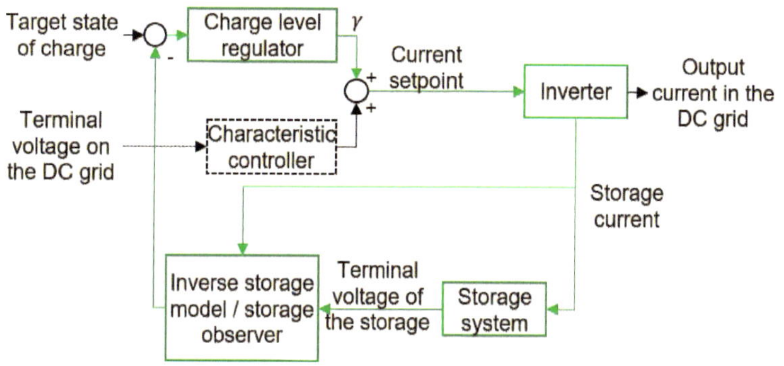


**Figure 6.16** Basic structure of the storage control with shifting of the control characteristic curve

If a simple proportional element is used for the state-of-charge controller, the curves for the output impedance $Z_{Prs}$ and the closed state-of-charge controller $G_{SOC}$ (SOC = state-of-charge) of the prosumer shown in Figure 6.17 are obtained. Here, $G_{DR}$ represents the set characteristic curve slope and thus the virtual internal resistance. It can be seen that the characteristic curve control is not effective in the frequency range below $f_{GR}$. In comparison, the state-of-charge control can only achieve its control objective of maintaining the state-of-charge at $SOC_{set}$ below this frequency. The cut-off frequency $f_{GR}$ depends on the gain of the state-of-charge controller and the transfer of the current setpoint to the state-of-charge; the cut-off frequency $f_{GR}$ can be determined from this.

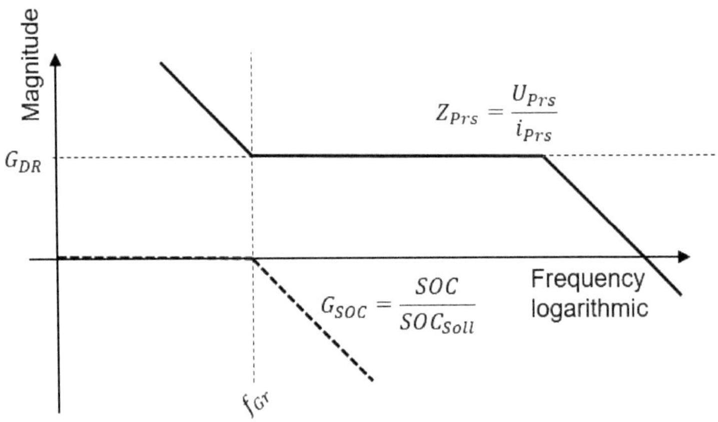


**Figure 6.17** Representation of closed loop state-of-charge control and output impedance of prosumer

The second possibility of a state-of-charge control consists in switching the control characteristics depending on the state-of-charge of the energy storage device. The basic concept is based on two different characteristic curves with different effects on the energy storage device.

There is one characteristic curve that leads to the predominant discharging of the energy storage device and a second characteristic curve that leads to the predominant charging of the energy storage device. The two characteristic curves are shown in Figure 6.18. The characteristic curves are switched over depending on the state-of-charge. The switchover can either take place in an embedded controller in the DC/DC converter or after being transmitted to the DC/DC converter by a higher-level controller.

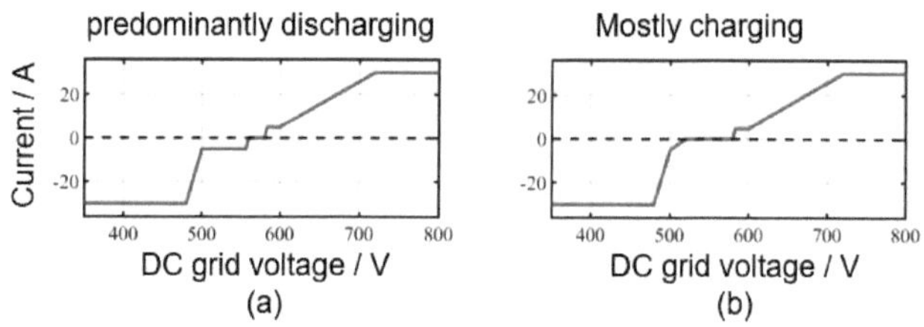


**Figure 6.18** Switching of two characteristic curves. Left: predominantly charging characteristic, right: predominantly discharging characteristic

If the discharge characteristic is selected at the current time, the energy storage device releases more energy into the DC grid. If the storage reaches a lower state-of-charge threshold, the system switches to the charging characteristic. The charging characteristic can be designed in such a way that charging is mainly performed by the braking energies occurring in the DC grid. The energy storage device is now charged until it reaches an upper state-of-charge threshold. Now the discharge characteristic is switched on again. This results in a two-point controller behavior for the secondary control target. The advantage of this method is that it can be used even if the load profile is completely unknown. This principle corresponds to the practical implementation of the decentralized group control in the CNC machining center described in Section 6.4.

## Literature

[Ott2015] Kirill: Modelling and Measuring Complex Impedances of Power Electronic Converters for Stability Assessment of Low-Voltage DC-Grids. IEEE First International Conference on DC Microgrids, IEEE, Atlanta, 2015

[Ott2019] Ott, Leopold: Modellbasierte Stabilitätsuntersuchungen in Niederspannungs-DC-Verteilnetzen. Dr. Hut - Verlag, Erlangen, Nürnberg, 2019

[Hämm2015] Hämmerle, Florian; Sandler, Steve; Schuster, Tobias: Bode 100 - Application Note: Traditional and Non-Invasive Stability Measurements. OMICRON Lab-V2.0, Smart Measurement Solutions, Klaus,2015, https://www.all-electronics.de/wp-content/uploads/2015/11/1-App_Note_Traditional_NoninvasiveStability_V2_0.pdf (access on 27 Mar 2018)

[Männ2019] Männel, Alexander; Tappe, Svenja; Knöchelmann, Elias; Ortmaier, Tobias: Investigation on an AC Grid Failure Handling of Industrial DC Micro Grids with an Energy Storage. 2019 IEEE International Conference on Industrial Technology (ICIT), IEEE, Melbourne, 2019

# 7 DC Grid Protection

Hartwig Stammberger
Johann Austermann
Olaf Grünberg
Frank Jungnickel
Wolfgang Veit
Wolfgang Hauer

## 7.1 Safety Requirements

For **system safety** during operation, the grounding concept and the grid design as described in Chapter 5.4 have to be observed to control short circuits or ground faults on the DC side. In the case of a grid connection with a grounded neutral point, protection must therefore be provided in both poles. When feeding with an open central point (analogous to the IT system), either only the positive or only the negative pole can be protected in the entire DC system, since an initial grounding fault is not yet a short circuit. Insulation monitoring must be provided to detect this grounding fault.

In the event of a fault, it must be possible to safely isolate the faulty DC sector from the system without affecting other DC sectors. In the event of a short circuit, a disconnection criterion is that the permissible temperature of the wire insulation is not exceeded.

For **personal safety**, grounding, connection technology and in particular the five safety rules in electrical installations (DIN VDE 0105-100) must be observed; this is explained in more detail in Section 7.3.

#### Physical Fundamentals of Faults in DC Grids

To understand the fundamental differences in the design and operation of protective devices for AC and DC grids, the characteristics of short circuits in DC grids are first considered. From the electrical behavior of arcs in air, the temporal development of DC short-circuit currents can be inferred. This results in requirements that must be considered in the design of DC breakers.

If a serious fault situation (e.g., line break, component fault, short circuit) occurs within the DC sectors or on the DC grid, an arc will form at the fault location similar to the situation in AC grids. In AC grids, the dominant inductances of lines, transformers and components in the system damp the current rise, so that a few milliseconds are sufficient to extinguish the arc and prevent damage to the grid.

The zero-crossing of the mains voltage following the fault also ensures that the fault current drops again instantaneously. This opens up a number of options for taking over and extinguishing the arc in protective devices, which can take effect within a few milliseconds. These include the mechanically operated extension of the arc, the splitting of the arc via several quenching plates, and the structural influencing of the gas ion currents.

In DC grids, the inverter capacitors present in all DC sectors determine the behavior in the event of a fault. The capacitances installed on the bus side add up to several tens of microfarads. In addition, larger storage capacitors can be connected directly to the DC grid for grid maintenance and grid support. The charge carriers stored there cause the short-circuit current to rise sharply. This is only counteracted by the comparatively low line inductances. Very high fault currents of several tens of kiloamperes develop within a very short time. This could damage the system or reduce the bus voltage to such an extent that the converters of the system components switch off.

For the design of suitable protective devices, it is initially useful to take a closer look at the temporal development of a fault situation in DC grids. The design depends on a realistic estimation of the occurring short-circuit resistance. As described above, an arc will develop in the majority of cases. For small currents, the gas in the arc is not fully ionized, so the number of charge carriers and thus the conductivity depend on the current strength. As soon as the gas is largely ionized at higher current strengths, a relatively constant arc resistance of a few milliohms is established.

A fault with a constant short-circuits resistance can be considered as a special case, however, this case would presuppose a fixed connection of the contacts involved. Such a case occurs in real plant grids only in the case of faulty installations. The characteristic dependence of the arc resistance on the short-circuit current at currents below a few hundred amperes is thus not realistically represented.

The resistance characteristic of the arc depends primarily on the electrode materials as well as on the length of the arc. Figure 7.1 shows the current dependence of the arc resistance for some classical arc models. The curves according to Paukert [Pauk1993] and Nottingham [Nott1923] show exemplarily the results of extensive evaluations of measurements by different authors. In Paukert's curves, the dependence of the arc resistance on the arc length $L_{arc}$ can be clearly seen. The relationship derived by Nottingham for short arcs of 1–10 mm length fits comparatively well with Paukert's curve with the arc length of one millimeter. In both cases, the difference to the special case of constant arc resistance is significant, especially for low fault currents.

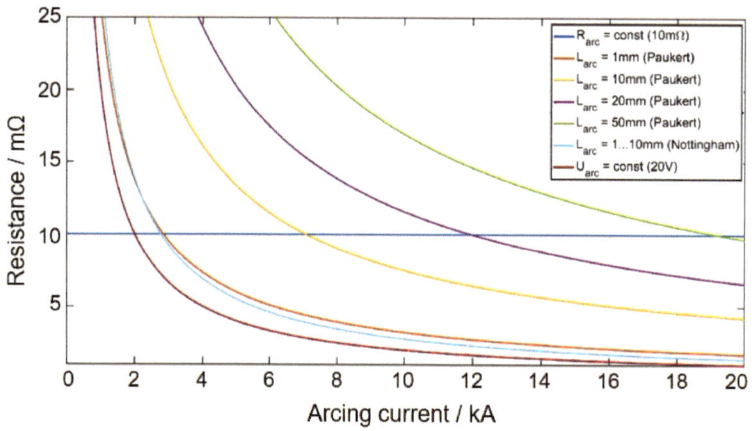


**Figure 7.1** Arc resistance as a function of arc current

In addition to the models mentioned, the curve for a constant arc voltage of 20 V is given. It shows the typical arc characteristic and can be interpreted as a lower limit function for minimum arc lengths. The assumption of a constant arc voltage for very small arc lengths is physically justifiable in that the arc in this case consists only of the cathode and anode drop zones. The voltage drop across the intervening plasma column in fact only makes a significant contribution to the arc voltage for extremely small short-circuit currents. However, the sum of the anode and cathode drop voltage is largely constant and depends only on the electrode material. In [Igna1997], for example, values of 23.5 V are given for copper and 26.5 V for steel. A detailed discussion of the arc problem can be found in [Amme2010].

Figure 7.2 shows the course of simulated short-circuit currents on the central DC grid of an industrial plant for the above short-circuit models. In each case, a +/- fault at 0.6 s is considered in the body-in-white cell of *Daimler* described in Chapter 9. In addition, the case of a completely resistance-free short-circuit ($R_{arc}$ = 0 mΩ) was included in the representation.

Common to all models is an extreme increase in current intensity to values >30 kA already below one millisecond. The high prospective fault currents result from the simultaneous discharge of all capacitors into the fault location on the DC grid.

After the capacitors are discharged, the fault currents approach steady-state values around three kiloamperes, given by the ohmic losses of the semiconductors, the filters, and the cable and connection technology in the system. In between, the currents of the "short" arcs as well as the resistance less short circuit drop to values around zero amperes or even below. The comparison with the resistance curve in Figure 7.1 indicates that the dynamic of the process leads to a tendency of the system to self-oscillate due to the temporarily very high currents. This tendency is largely suppressed in the case of "long" arcs, but especially in the case of finite (constant) short-circuit resistance. The effect depends on the given system but can result in self-extinction of the arc within 300–500 ms even in the case of DC short-circuits – although this is unreliable because of the steep rise in current to several tens of kiloamperes after the short-circuit has occurred.

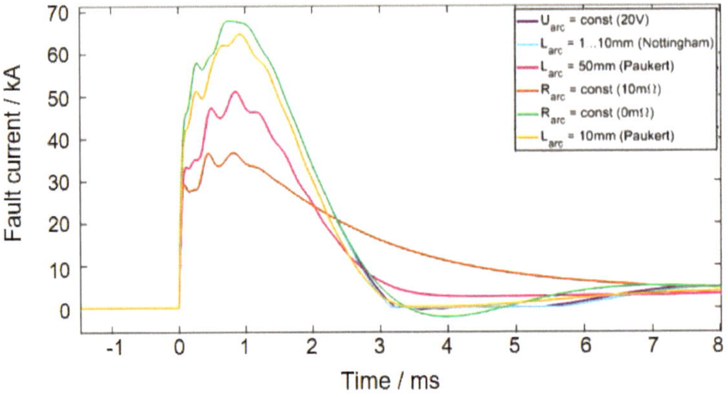


**Figure 7.2** Development of the fault current in time (overview)

The corresponding bus voltage curve (measured at the AIC) is shown in Figure 7.3. Values around 520 V (lower limit for steady state undervoltage) are already undercut at 40 µs. However, after 60 µs (400 V) at the latest, all converter-operated system parts are threatened with transition to uncontrolled operation or activation of self-protection. The simple arc model with $U_{arc}$ = const (20 V) also provides a suitable lower limit case for the design of the circuit breakers.

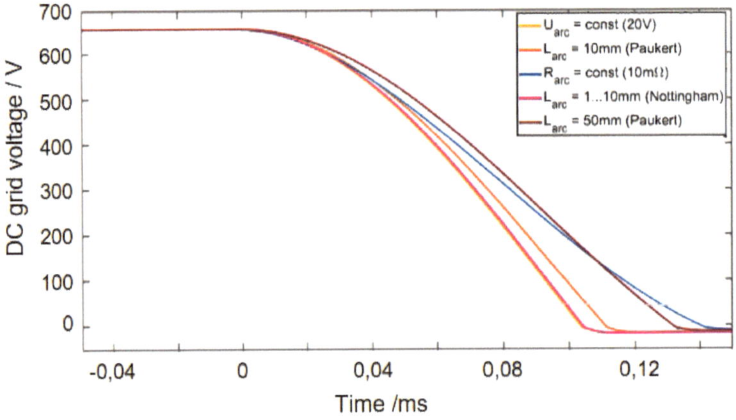


**Figure 7.3** Development of the fault current in the AIC DC sector

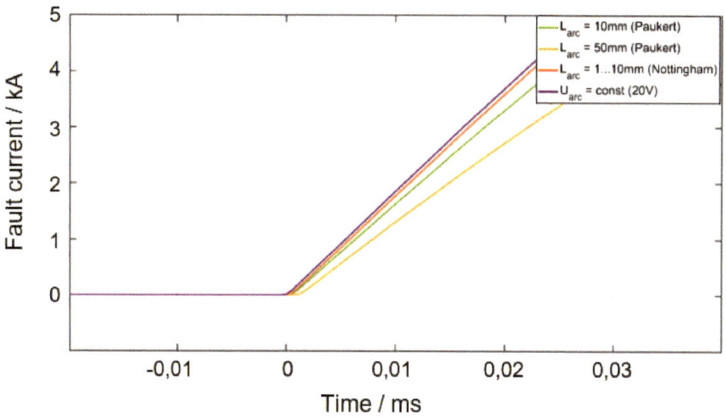


**Figure 7.4** Course of the bus voltage over time measured at the AIC

The fault current on the DC grid shown in Figure 7.2 is made up of the contributions of the individual DC sectors. These current contributions are of interest for the design of the corresponding circuit breakers. From the course of the fault current in the AIC DC sector (Figure 7.4), it can be seen that again the "short" arcs and the simple arc model place the highest demands on the speed of the protection. Short-circuit currents of almost two kiloamperes are reached already after 10 µs. Depending on the type of fault and the DC sector, di/dt values of up to 200 A/µs and $I^2t$ values of 10–20 $A^2s$ must be expected.

This results in two possible configurations for the design of circuit breakers, which are described in more detail in the following chapters:

1. Semiconductor breaker

   Semiconductor devices (IGBTs) are capable of interrupting the short-circuit current within a microsecond. The challenge here is to detect the short circuit quickly and inexpensively so that the short circuit current can be cut off within a total of 10–20 µs.

2. Hybrid breaker

   In order to circumvent the disadvantage of the permanently occurring power loss of IGBTs in the power path, the detachable contact in the breaker is designed mechanically. Only in the event of a short-circuit do the semiconductor components – which are arranged in parallel – take over the commutation current, so that the arc that occurs upon contact opening is extinguished quickly since the current commutates to the lower-resistive semiconductor path. The process is slower overall (several hundred microseconds), so that the combination of mechanical and semiconductor breaker must be connected in series with an inductance, which significantly reduces the current rise in the DC sector described above.

## 7.2 Plant Safety

The grid is divided into so-called DC sectors, as shown in Chapter 5.1. The loads of the zones should be distributed as evenly as possible, i.e., there should be no single dominant load, so that effects on neighboring zones are minimized in the event of a fault. It shall be ensured that during a shutdown, the voltage in the other DC sectors remains within the permitted voltage range (see Section 5.3).

A circuit breaker is required at the input of each DC sector to handle short circuits, overloads, and in some designs, ground faults. In the event of a fault, the short-circuit current rises very steeply because of the many capacitors and storage devices in the grid - up to several 10 A/µs. Therefore, a very fast detection and disconnection of the short circuit is necessary in order not to overload the equipment and the semiconductors in the converters. Such a protection cannot be achieved with mechanical switching devices, or only with very huge efforts. Due to the longer switch-off time of mechanical switching devices, compliance with the voltage band (Chapter 5.3) can generally no longer be guaranteed in the event of a short circuit. Therefore, the use of power electronic switching devices is recommended, either as hybrid or semiconductor breakers.

Both circuit breaker types switch off the short-circuit current in such a rapid way that thermal effects on the supply lines to the DC sector or on the fault location do not play a role in the dimensioning of the circuit breakers. Purely mechanical state-of-the-art circuit breakers with a trip time of several milliseconds are not suitable for the DC grid, since they cannot prevent the impairment of other DC sectors in the event of a fault.

For the protection of DC sectors connected in parallel, it is advantageous to use circuit breakers with the same operating principle (i.e., hybrid breakers or semiconductor breakers) in order to achieve the best possible selectivity, since the trip times differ significantly in the event of a fault. Mixed operation, e.g., operation of many small DC sectors with hybrid breakers and one large DC sector with fully electronic breakers, is possible in principle, but must be checked in each individual case.

#### Hybrid Breaker

Hybrid breakers combine the advantages of the low power dissipation of metallic contacts for permanent current conduction with the speed of disconnection in the event of a short circuit by means of power semiconductors that can be switched off (e.g., IGBTs).

Hybrid circuit breakers consist of a parallel connection of a low-impedance electromechanical contact (bypass relay) with a semiconducting switching component, as schematically shown in Figure 7.5.

The block diagram of the hybrid breaker used in the DC-INDUSTRIE project is shown in Figure 7.5. From left to right, the all-pole maintenance breaker, the air coil for limiting the current rise, and the actual hybrid breaker are shown. The hybrid breaker itself consists of the parallel connection of a **mechanical contact** (bypass relay), the bidirectional **power semiconductor** module and a **varistor.** The **shunt** is used for current measurement - but of course other methods for current measurement are possible. **Isolation contacts** isolate both poles from current after the current is switched off (by the power contacts) and ensure the isolation requirements.

Figure 7.6 shows the principle of the current paths when the hybrid breaker is switched off.

a) In normal operation, the mechanical contact alone carries the current.

b) In the event of a fault, the semiconductor is switched on additionally and most of the current initially flows through the mechanical contact because its resistance is lower.

c) After (rapid) opening of the mechanical contact, the power semiconductor carries the current alone.

d) When the opening distance of the mechanical contact is sufficiently large, the power semiconductor is switched off and the varistor takes over the current flow (not shown here, cf. Figure 7.5 for the varistor).

e) Finally, the isolation relay is opened without current.

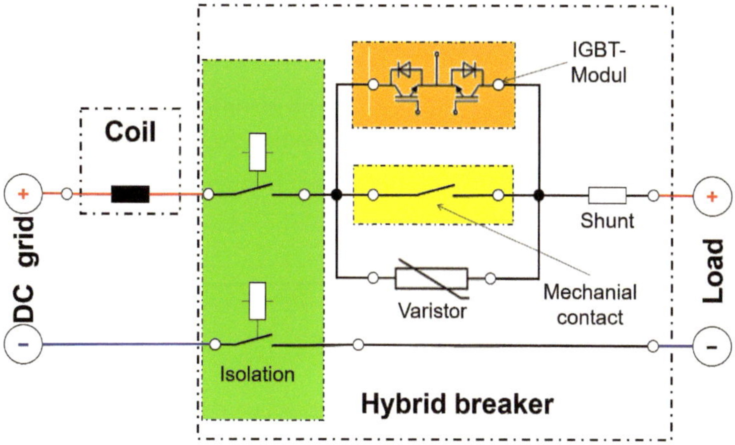


**Figure 7.5** Schematic diagram of a hybrid breaker [Stam2019]

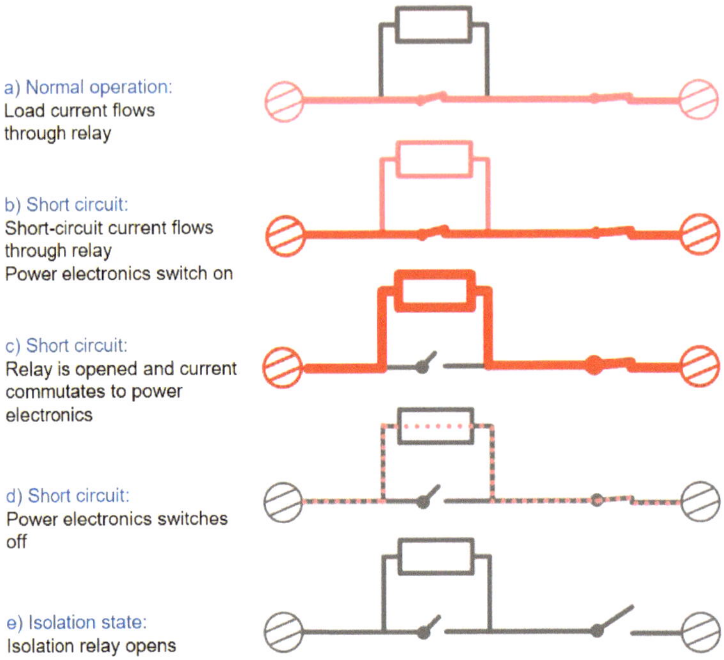


**Figure 7.6** Principle of current conduction when switching off the hybrid breaker [Aska2018]

When **switching off** – either manually or when the current-time curve is outside the set tripping characteristic – the IGBT module is switched on and the mechanical contact is opened via a fast actuator. The processes can be understood from the schematic diagram in Figure 7.7 (whose time and current axes are not to scale):

1. At time $t_0$ the fault occurs, the current starts to rise and continues to flow first via the mechanical contact.
2. At time $t_1$ the fault is detected, and the power semiconductor module is switched through. Because of the lower resistance of the mechanical contact compared to the on resistance of the semiconductor module, the current continues to flow through the mechanical contact (subfigure b)).
3. At time $t_2$, the mechanical contact opens, and an arc is generated whose minimum arc voltage of typically 20 V is significantly higher than the forward voltage of the switched-on power semiconductor. Therefore, the current commutates from the arc to the power semiconductor – for fast commutation, care must be taken to ensure a low-inductance structure between the mechanical contact and the power semiconductor.
4. The commutation is completed at time $t_3$, the current now flows completely via the power semiconductor, see subfigure c). Meanwhile, the mechanical contact continues to open until it has reached a sufficiently large opening distance to withstand the recurrent line voltage.

5. At this time $t_4$, the power semiconductor is blocked and the energy in the system is led to zero via the varistor (subfigure d).
6. When the current is zero (time $t_5$), the isolation relay is opened without current.

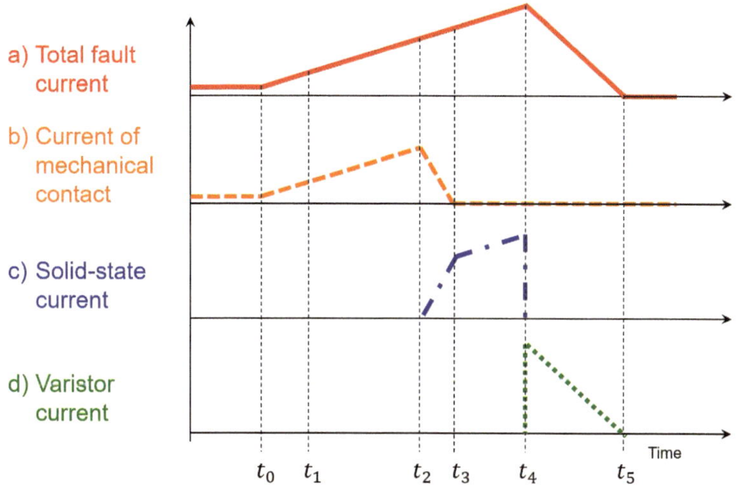


**Figure 7.7** Principle current waveforms when a hybrid breaker is turned off [Aska2018]

The time to first open the contact is typically less than 400 μs. To achieve this short time, a fast actuator and low inertia of the contact system are necessary. The parallel varistor limits the voltage and dissipates the remaining energy in the circuit until the current is extinguished. The magnitude of the required opening distance of the mechanical contact and thus the turn-off time of the power semiconductor as well as the joule energy $\int i^2 dt$ at turn-off is determined by the response voltage of the varistor.

The inductor in series with the breaker serves to limit the rise of the current in the short circuit via its inductance $L$. In DC grids, current rise times of several 10 A/μs are achieved due to the many capacitors, which - without limitation - would lead to a high load on the mechanical contact.

In order to be able to implement the hybrid breaker technology in an economically meaningful way, high requirements are placed on the bypass relay with regard to the opening speed, especially in the event of a short circuit. In this case, the power semiconductor is switched on and the bypass relay is opened at the same time. Due to the mechanical inertia of the bypass relay (armature), the contact separation does not take place immediately. Only after a delay time, which mainly depends on the mechanical design, the bypass relay opens. During this time, the current increases. Since the contact resistance of the electrical contact is about 100 times smaller than the resistance of the power electronics, the short-circuit current flows mainly through the bypass relay. Only after contact separation does the current fully commutate to the power semiconductor. The amount of current that the power electronics must disconnect thus depends on the opening speed of the relay. The faster it opens, the smaller the maximum current carrying capacity of the power electronics can be selected, since the current cannot rise as high. However, since in DC grids the current rise in the event of a short circuit can reach very high values (>10 A/µs) due to the capacitive nature of the grid, a so-called "time-buyer coil" is also installed in the current path to be protected. This air coil limits the current rise so that the current does not exceed the maximum current carrying capacity of the power electronics after the contacts are opened.

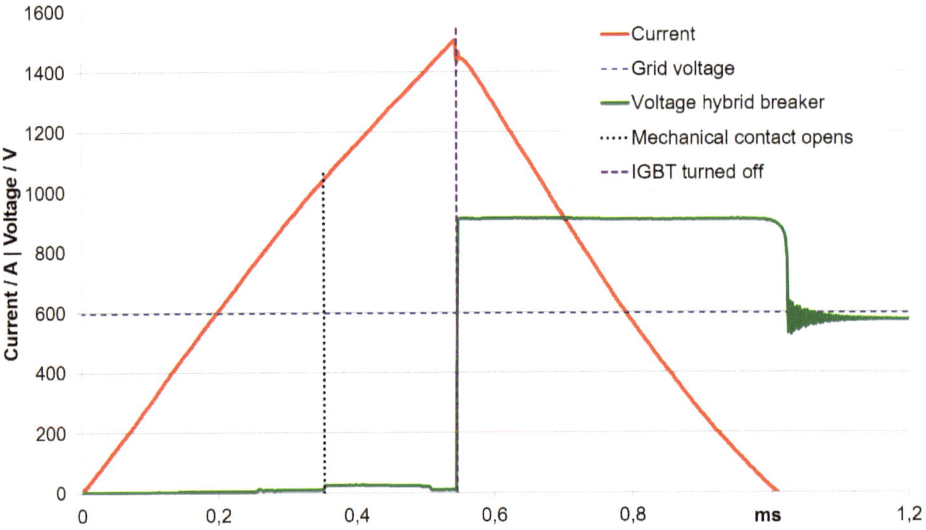


**Figure 7.8** Shutdown oscillogram for a hybrid breaker, 600 V, 12 kA prospective [Haue2018]

Figure 7.8 shows an example of the switch-off oscillogram of a hybrid breaker with 44 A rated current at 600 V for a prospective current of 12 kA. After about 350 µs (dotted line), the mechanical breaker opens (bypass) and the current commutates to the power semiconductor; after another 200 µs or so, the mechanical contact is open far enough and the IGBT is switched off. The voltage at the hybrid breaker jumps to the approximately 900 V response voltage of the varistor, which carries the current from its maximum value (approximately 1500 A) to zero. After about a millisecond, the current is off and the voltage across the breaker is equal to the voltage across the DC grid.

The let-through energy $\int i^2 dt$ of the circuit shown here at 600 V DC line voltage is about 750 $A^2s$ – this is nearly three orders of magnitude smaller than the approximately 600 $kA^2s$ of a compact mechanical circuit breaker. This clearly demonstrates the superior limitation of fault current and energy provided by the use of power semiconductors.

### Semiconductor Breaker

The rapid advancement of power semiconductors to lower power dissipation, higher overload capability and greater dielectric strength means that semiconductors are increasingly being used in switching applications.

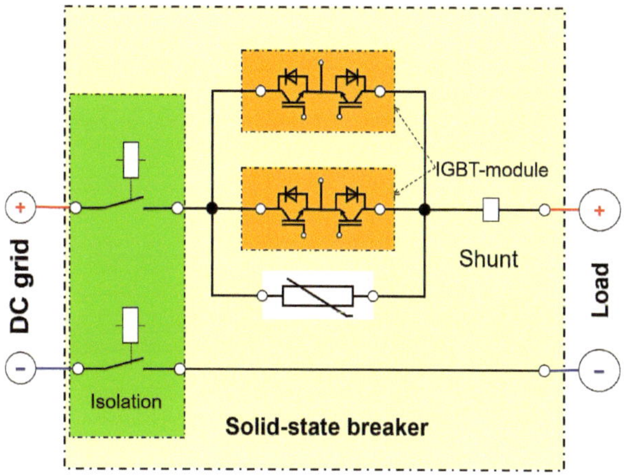


**Figure 7.9**
Schematic diagram semiconductor breaker [Stam2019]

The design of semiconductor breakers according to [Stam2019] as shown in Figure 7.9 differs from the hybrid breaker in that there is no mechanical switching contact (except for the current-less opening insulation contact) and no coil. As a result, once the fault is detected, the current is interrupted very quickly – in the range of a few 10 µs – and the load is disconnected from the mains. The super-fast disconnection has the advantage that the voltage in the DC grid is only marginally affected, even in the event of a short-circuit. However, nominal current flows permanently in the semiconductor breakers. These must therefore be cooled at greater expense than would be the case with hybrid breakers. For smaller power dissipation and for larger rated currents, several semiconductor modules can be connected in parallel.

For self-protection of the device, however, it always has to be ensured that the device switches off in a few µs, since the maximum current-carrying capacity of the semiconductors is limited. The energy in the event of a fault $\int i^2 dt$ is several orders of magnitude smaller than is the case with classical mechanical DC switches. Moreover, a coil to limit the current rise in case of a fault can be omitted. In addition to the omitted power dissipation, this has the advantage that the risk of excitation of current harmonics in the overall system is reduced.

### Comparison of Switching Principles (Hybrid and Semiconductor Breakers)

Compared to conventional mechanical breaking devices, hybrid and solid-state breakers switch off the current much faster. This results in significantly lower energy in the event of a fault. Figure 7.10 shows principle current-time curves in a short circuit: a circuit breaker (red curve) limits the current less clearly than the miniature circuit breakers usually used in households (blue curve) due to its design. Hybrid and semiconductor breakers have a significantly shorter breaking time, as can be seen from the green curve in Figure 7.10 and the comparison between hybrid and semiconductor breakers in Figure 7.11.

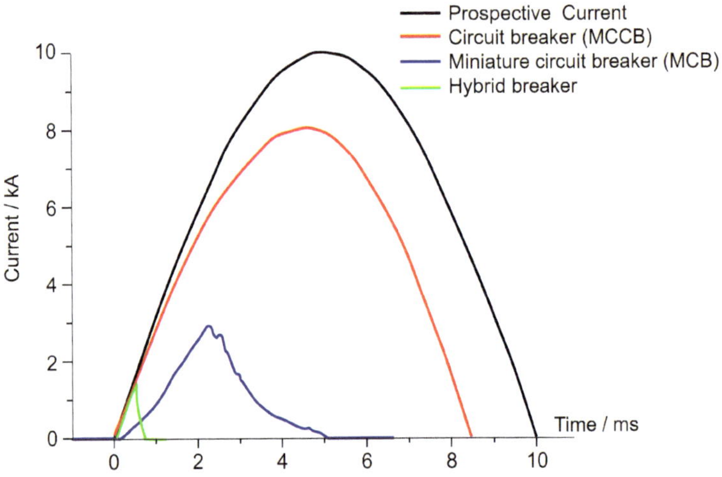


**Figure 7.10** Principle current characteristics when switching off due to a short circuit for different breaker types [Haue2019]

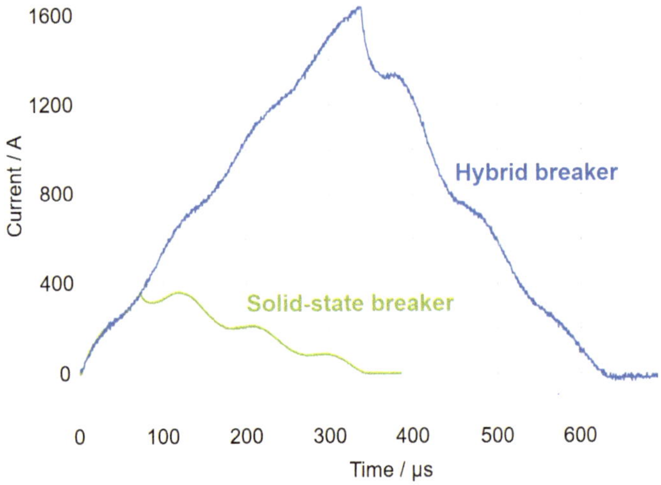


**Figure 7.11** Comparison of current waveform for hybrid and semiconductor breakers during short circuit [Haue2019]

### Overload Protection for Lines and Devices

For device and line protection, it is necessary to know the current-time characteristics and, if applicable, the $\int i^2 dt$-values of the DC sectors. In contrast to AC grids, the power supply in DC grids is always provided by semiconductor components (e.g., diodes and/or IGBTs in the supply devices), which have much lower overload capability compared to AC components (e.g., transformers) and whose protection therefore places high demands on the dynamics of the protection devices. A high current load on the semiconductors in the supply devices can occur, for example, when the mains returns after brief power interruptions. Conceptually, device protection is ensured by current measurement and a $\int i^2 dt$ model in the protection devices of the Smart Breakers. The $\int i^2 dt$ model must be matched to the supply devices and to the devices in the DC sector that is to be protected. In the case of overload, the shutdown criterion is that the permissible operating temperature of the cable insulation is not exceeded.

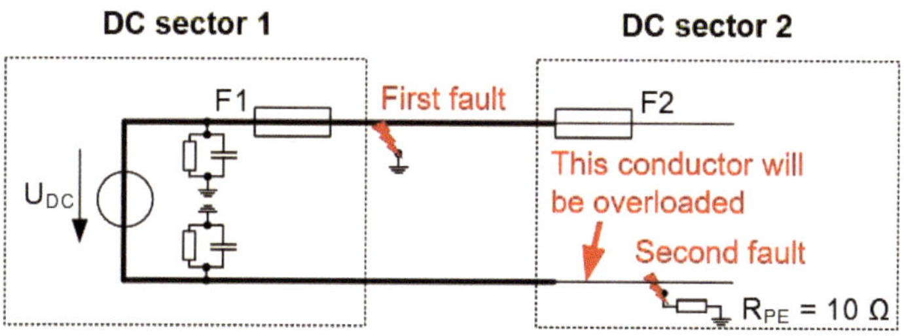


**Figure 7.12** Conductor protection in case of two ground faults in different DC sectors

State-of-the-art line protection must also be provided within the DC sectors. This can be done, for example, by means of protective devices or short-circuit-proof cable routing.

In the case of the IT system, protection devices are only provided for the DC+ line (see Chapter 5.4): In the event of a grounding fault in the positive or negative DC pole, respectively, the DC grid can continue to operate for a short time; however, special attention must be paid to line protection with this type of grounding, as explained in the following.

Figure 7.12 shows that in DC sector 2 the cable cross-section is reduced. In the event of a second fault in this DC sector with reduced cross-section, the line protection could be overloaded without further protective measures if the resistance is sufficiently small. Therefore, in this case of reduced cross-section in a DC sector, the negative pole must also be protected against overload.

### Selectivity for Targeted Protection

Selectivity is the ability of a plant to shut down only those of its parts that are directly affected in the event of a fault, ensuring continued operation of the fault-free parts as much as possible. When there are many energy storage devices in a plant, ensuring selectivity becomes more challenging than in a traditional linear structure with a centralized power supply, since in the event of a fault all the storage devices will attempt to discharge into the fault location. For selectivity, i.e., maintaining operation in DC sectors that are not affected by the fault, it is therefore necessary to ensure that the circuit breakers of these unaffected DC sectors do not trip when a fault current is supplied from storages in their DC sector into the fault branch. This can be done, for example, by detecting the direction of the current in the Smart Breakers. When current flows from the DC sectors towards the DC grid, the tripping of the circuit breaker should be delayed giving the breaker of the faulted sector time to trip the fault.

Figure 7.13 shows an example of a short circuit (red arrow) in DC sector 2.1. Only the breaker of this DC sector should then switch off, while the breakers of the parallel DC sectors (yellow) as well as the neighboring DC sectors (light green) and those of the higher hierarchy levels (blue) should remain switched on.

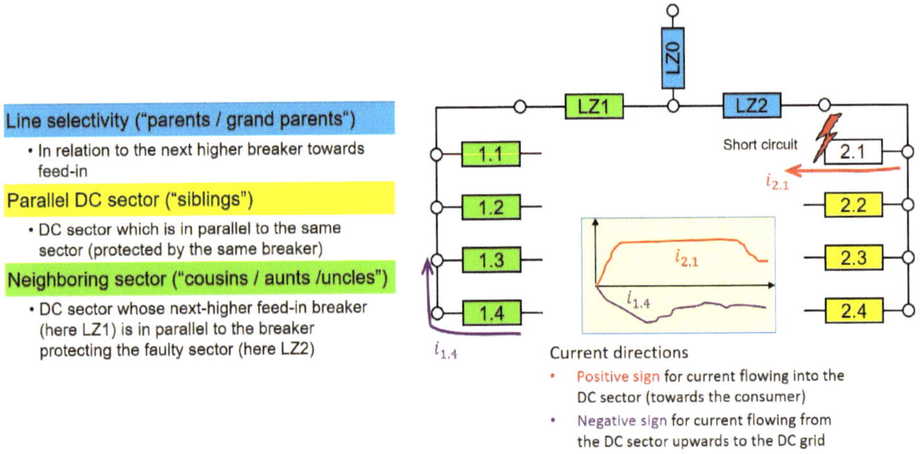


**Figure 7.13** Definitions of selectivity

Another criterion for ensuring selectivity is compliance with the voltage band (see Chapter 5.3). If the fault takes too long to switch off, the voltage in the DC grid continues to drop – the Smart Breakers would be switched off via the voltage monitoring system if the voltage falls below the undervoltage limit.

## ■ 7.3 Personal Safety

According to DIN VDE 0105-100, personal safety is based on the five safety rules, which are to be applied exclusively by qualified electricians and persons trained in electrical engineering. They serve as a measure for protecting against electric shock. In companies, occupational safety is responsible for ensuring that correct and prudent action is taken. Thus, in principle, a safety-related set of rules is in place.

Generally, on the DC-side, a high-impedance center grounding as has been described in Section 5.4, is preferred, because it is advantageous in terms of personal safety, especially when being fed from an AC grid with an ungrounded neutral point. Detailed studies covering - for example - DC supply grids for telecommunications equipment and data centers, highlight the advantages of such grounding systems. In Hirose et al [Hiro2011] - as an example - reference is made to the publication IEC/TR 60479-5, which explains the effects of currents on the human body and gives recommendations for limits of direct currents and low-frequency alternating currents. The graph in Fig. 7.14 (from [Hiro2011]) evaluates the potential harmfulness of direct currents and low-frequency alternating currents when the current takes the path from both hands to both feet:

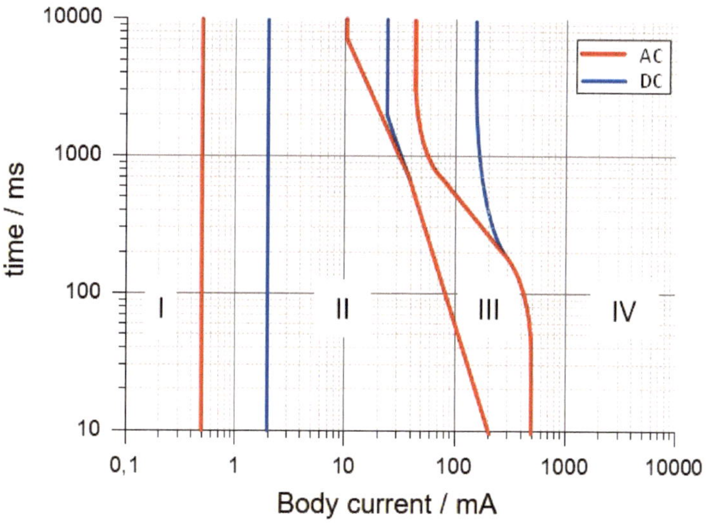


**Figure 7.14** Definition of critical ranges for height and duration of body current flow at DC and low-frequency AC voltage according to IEC 60479-1 (current flow from both hands to both feet; ascending direction of current flow at DC voltage)

The ranges shown in Figure 7.14 can be classified as follows with respect to their harmfulness to the human body:

- Range I: Normally no physiological perceptibility
- Range II: Perceptibility threshold exceeded, but normally no dangerous physiological effects or occurrence of pain
- Range III: Occurrence of reversible effects, such as muscle contractions and respiratory distress, highly probable
- Range IV: Very high probability of occurrence of irreversible physiological effects and ventricular fibrillation

In general, it can be seen in Figure 7.14 that the thresholds for the individual ranges are higher for longer durations of current flow in the case of DC voltage than for AC voltage. When using a high-impedance center grounding for the DC mains system, it is easily possible to limit the continuous DC current through the human body to harmless values in case of accidental contact with a DC mains pole, if the grounding resistances are correctly dimensioned. Assuming a body resistance of 1.5 kΩ and a maximum DC mains voltage of 800 V symmetrically distributed around the grounding potential, a cumulative ground resistance of at least 14.5 kΩ is tolerable for the DC mains pole concerned in order to limit the continuous DC current to 25 mA and thus not to leave the still uncritical range II.

With regard to the "plugging connection" of two electrical conductors, a distinction must be made between plug-in devices and plug-in connectors. Plugs and sockets are intended for use by laypersons, connectors for use by qualified personnel. This is (usually) accompanied by the functional difference that plugs and sockets may also be plugged or unplugged under load. In contrast, this is generally not the case with connectors. The above rule applies to connectors, i.e., they must be switched load-free both before being plugged in and before being unplugged. This technical difference is reflected, among other things, in the higher technical effort required for plug-in devices compared to plug-in connectors, even though the performance data are identical. For example, a plug-in connector with the dimensions of a standard household appliance plug-in device for 16 A would be used in industrial applications up to 45 A. Alternatively, technical measures would be found in the products to allow plugging and unplugging under load, with corresponding costs, which would be contrary to industrial production use. Therefore, in the case of DC grid, it is advisable to rely on connectors and to ensure that plugging and unplugging under load is

a) not the intended application,

b) in the event of a fault, will result in only limited damage to the connector,

c) in the event of a fault, does not lead to personal injury.

The related results from the DC-INDUSTRIE research project suggest an extension of the standard for DC connectors. In addition to defining general connection requirements, the unauthorized disconnection of plug and socket must be prevented. In the case of simple unlocking (level 1), unintentional pulling of the connectors is prevented. Extended unlocking (e.g., safety clip) enables deliberate unlocking by hand (stage 2) or with a standard specialist tool (stage 3). Finally, stage 4 provides for a further extension of the unlocking (e.g., lockable safety clip), for example by means of a key or a special tool, or in the case of automatic locking, the opening of the connector is authorized by a defined state.

For future connectivity, the interchangeability of the connection technology in use must be fulfilled (definition according to IEC in IEV 581-24-03). This means: the form, dimensions and function as well as the technical properties of the connection technology on the drives, the switchgear, the supply devices and the infrastructure components of a power and installation class (hybrid or pure DC power) from different sources must be identical. In other words, it must be possible to replace a connector from manufacturer A – regardless of whether it can be freely assembled, has preassembled cables or is integrated in the device – with a connector from manufacturer B, while retaining the functionality of the overall connection in all its technical properties (electrical, mechanical and climatic). Figure 7.15 gives an overview of the connection technology.

Particularly regarding infrastructure connectors for hybrid lines, the aim is to keep the number of variants low by means of a scalable or modular concept. The infrastructure connectors are used where power (main and auxiliary power), Ethernet-based data and signals (such as STO – Safe Torque Off) are distributed outside the control cabinet or on the control cabinet wall.

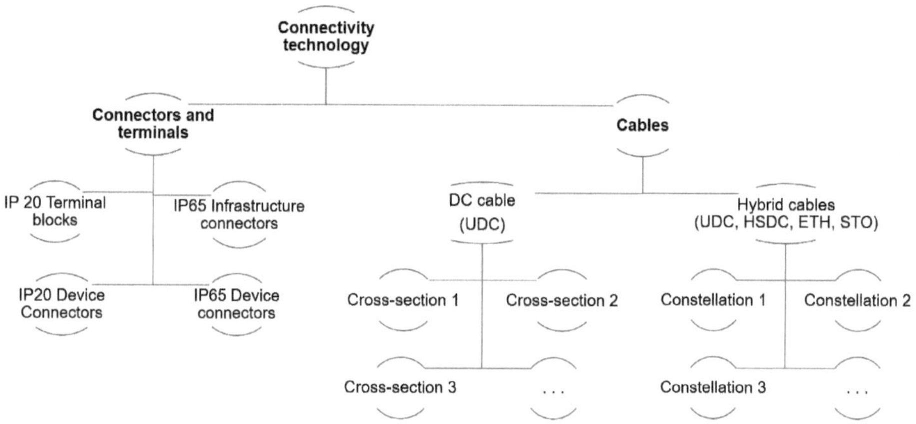


**Figure 7.15** Overview of connection technology for connectors, terminals, and cables [Grün2019]

Depending on the grounding reference, operating voltages may have to be redefined, since operating voltages are currently only defined for "pure" DC or AC operation. There is no definition for voltages with high common mode content (relevant for AIC fed DC grids).

The charged capacitances in the devices or DC sectors represent a potential hazard. Thus, the discharge times are a special feature of DC grids and should be given special attention. The solution for this can be realized via the discharge function, e.g., in the Smart Breaker (see Chapter 5.5).

## Literature

[Stam2019] Stammberger, Hartwig; Köpf, Hendrik: DC-Netz schützen und sicher schalten. Conference "DC 2019 – Industrielle Gleichstromnetze", Technische Hochschule Ostwestfalen-Lippe, Lemgo, 2019

[Haue2018] Hauer, Wolfgang; Stammberger, Hartwig; Grünberg, Olaf; Wieghofer, Moritz; Austermann, Johann; Diendorfer, Christoph: Kurzschlussprüfungen für DC-Abzweige. Fachhochschule Oberösterreich, Wels, 2018

[Aska2018] Askan, Kenan; Bartonek, Michael; Sobe, Klaus: Design and Development of an Integrated Power Module Used in Low Voltage DC/AC Hybrid Circuit Breaker. PCIM Europe 2018, VDE, Nuremberg, 2018

[Amme2010] Ammerman, Ravel; Gammon, Tammy; Sen, Pankay; Nelson, John: DC-Arc Models and Incident-Energy Calculation. IEEE Transactions on Industry Applications, IEEEE, Los Alamitos, 2010

[Grün2019] Grünberg, Olaf: Verbindungstechnik in der Infrastruktur von DC-Netzen. Conference "DC 2019 – Industrielle Gleichstromnetze", TH OWL, Lemgo, 2019

[Pauk1993] Paukert, J: The arc voltage and arc resistance of LV fault arcs, Institute of Electrical Apparatus Lodz, Lodz, 1993

[Nott1923] Nottingham, Wayne: A new equation for the static characteristic of the normal electric arc. *Journal of the American Institute of Electrical Engineers*, IEEE, Los Alamitos, 1923

[Igna1997] Ignatko, V.P.: Electric characteristics of ac open heavy-current arcs. Proc. 3rd Int. Symp. Switching Arc Phenom., 1977

[Haue2019] Hauer, Wolfgang, Bartonek, Michael: Power Electronics & DC-Grid Solutions. interne Präsentation, Eaton, Wien, 2019

[Hiro2011] Hirose, Keiichi; Tanaka, Toshimitsu; Babasaki, Tadatoshi; Person, Sylvain; Foucault, Olivier; Sonnenberg, BJ; Szpek M.: Grounding concept considerations and recommendations for 400VDC distribution system. IEEE 33rd International Telecommunications Energy Conference (INTELEC), IEEE, Amsterdam, 2011

[IEC2007] IEC/TS 60479-1:2007: Effects of current on human beings and livestock – Part 1: General aspects. 2007

# 8 Planning and Design of a DC Factory

Darian Schaab
Alexander Maennel
Sebastian Weckmann

After the conceptual basics of the DC grid for a DC factory have been described in detail in the previous Chapters 5–7, this chapter addresses the design of a DC grid for a DC factory. Firstly, we present the relevant design objects of a DC factory. Then we highlight the differences to AC factories and enlarge upon in what way the established design procedures for AC grids have to be expanded in the context of designing a DC grid.

## 8.1 Relevant Design Objects in a DC Factory

The design objects of a factory comprise not only the equipment for the value-adding processes as well as the connecting logistics, but also the equipment for the peripheral, supporting processes, e.g., the ventilation for the factory buildings, but also the energy supply. While the value creation within the DC factory remains untouched, the equipment and systems must be adapted to the changed electrical energy supply. This means that the interface between the processes (e.g., transport, mechanical processing, ventilation) and the electrical energy supply system must be changed. Whereas the systems were previously connected to AC 400 V, in the future they will be serviced from the DC 650 V grid. The respective interfaces are located in the control cabinets of the plants. There, devices for rectification for DC intermediate circuits are omitted. Converters must be examined with respect to the requirements of the new supply and replaced if necessary. In addition, the protection technology must be redesigned, since the devices differ significantly from the previous AC devices. Background on this can be found in Chapter 7. There are different strategies for migrating an existing AC factory to a DC factory, depending on the life cycle of the factory itself and the equipment within the factory. If the new design of a factory – a "greenfield planning" – is carried out, the necessary interfaces in the plants as well as a DC protection concept should be provided right away. The economic potentials and characteristics of the DC factory are explained in Chapter 2.5. The rescheduling or partial modernization of a factory in which DC is to be partially implemented – a "brownfield planning" – requires the adaptation of the interfaces and the protection technology at selected plants and in certain areas. The investment costs rise gradually and can be planned strategically along the migration path.

In addition to the interfaces within the plants, the interface to the supply grid is of particular importance. For a complete DC plant, this is located at the grid connection point with the transformers present there. Here, the Active Infeed Converter (AIC) is installed directly on the secondary side of the grid-connection transformer. An additional isolating transformer is usually not necessary. Power factor correction equipment and some of the AC-side EMC filters can be omitted. If the DC grid is only used in part of a factory, an infeed point should be provided as an AIC with an isolating transformer or, in the case of simple installations, as an uncontrolled bridge rectifier. The latter only requires a line reactor instead of the isolating transformer. However, it does not offer the possibility of active grid management nor that of energy recovery. Further differences are described in Chapter 5.4.

A particular advantage of the DC grid and thus of the DC factory is the integration of electrical storage systems as well as the integration of decentralized energy generation into the factory's energy supply system. Figure 8.1 shows schematically how such a DC factory integrates storage systems and energy generators.

The storage units can be used in a decentralized way in the vicinity of the production plants, for example for peak load smoothing. Higher-level storage units for flexibilization and protection of the factory's energy supply can be connected at the highest power level of the DC grid. The grid management system to be designed actively regulates the use of the respective storage units depending on the intended grid function. Parallel to the storage units on the upper power level, different decentralized energy generation plants can be coupled to the DC grid. Here, the grid management offers the possibility of prioritization, i.e., controlling the use of energy sources depending on their costs. In addition, the transformability of the energy system is supported, since new energy generators can be easily integrated and expanded. Thus, future demands and cost situations can be addressed rapidly and efficiently.

The grid system dynamics in the distributed DC grid plays an important role in the stable and efficient operation of thc grid. First critcria can be found in the passivity of the participants (see Chapter 6.3). In addition, investigations in extended DC grids on the origin, effect and damping possibility of oscillations are necessary.

Overall, a DC factory with its design objects offers a modern form of energy distribution. By providing energy data and the ability to influence grid behavior, the DC grid supports smart grid transformation, the introduction of energy management solutions and "Industry 4.0" initiatives in companies. DC protection equipment –

the Smart Breakers – provide communication interfaces to transmit DC sector current and power flows to energy management systems. In addition, the Smart Breakers provide the ability to shut down the DC sectors. The decentralized group control of the grid management oversees the grid behavior by influencing the control characteristics. Depending on external influences, such as energy availability and costs, the operating point of the grid can be changed. Storage facilities and distributed power generation can be flexibly reprioritized and integrated in a modified way.

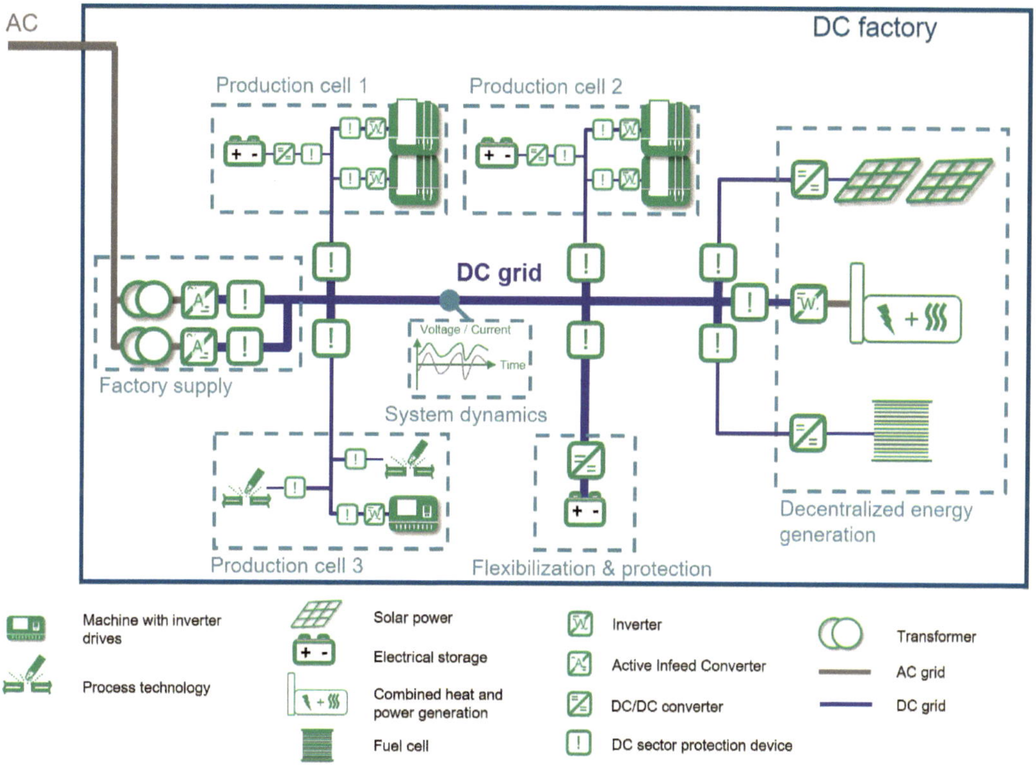


**Figure 8.1** Design objects in a direct current factory

## 8.2 Differences in the Design of DC and AC Grids

The design of DC and AC grids for an industrial plant or production facility differs regarding technical aspects and with respect to the functional design of the respective grids.

The main technical differences are presented in the system concept described in Chapter 5. The dimensioning of most of the operating equipment for DC systems is similar to the procedure undergone for project planning of AC systems, with the difference that the direction of energy flow is no longer unambiguously fixed from the higher hierarchical level to a lower level but can be reversed at any time.

A new operating device that has no direct equivalent in AC technology is the Smart Breaker. It makes it possible to precharge and selectively protect sections of the grid. These functions are a prerequisite for the operation of the DC grid. The precharging of grid sections must be considered in the project planning phase, as it must be dimensioned depending on the downstream capacity and the precharging technology. The precharging of grid sections is discussed in detail in Chapter 5.6.

Similarly, to achieve selectivity in a DC grid, the protection elements must meet higher requirements for short and adjustable disconnection characteristics because the DC grid is operated decoupled from the external grid. In the event of a fault, the inflow of energy is not only limited by the external grid impedance but depends on the dynamics of the power converters between the energy sources and the DC grid. At the first moment, the capacities installed in the grid alone feed into a fault location and, with low line inductance, cause high rates of current rise and a rapid drop in voltage. For selective disconnection of the fault, it is necessary to tune the characteristics of the Smart Breakers, see Chapter 5.5.

An AC power distribution system can be understood as a component of a historically developed power system. The electricity system of the Federal Republic of Germany is divided into three hierarchical levels: Transmission grid level, distribution grid level and private grid sections within buildings and/or empty lots. The project planning of grids or grid sections shows problems similar to those encountered in dimensioning transmission elements with regard to their current carrying capacity. Significant differences in the design result from the function of the respective hierarchy level: The transmission grid conducts energy over long distances and assumes system responsibility for balancing feed-in and withdrawal power in the grid. The distribution grid disperses energy from central feed-in points over wide areas to the connection point of the end consumer. This hierarchy level has the system responsibility for maintaining voltage within the specified tolerance band [Deut2005].

The distribution grid provides power to private grid sections within buildings or industrial plants. From the point of view of the electrical power system, these are the final consumers. This chapter is mainly concerned with the functional design of the DC grid, which - at the level of the private grid sections (Figure 8.2) - differs significantly from the project planning of an AC power distribution system.

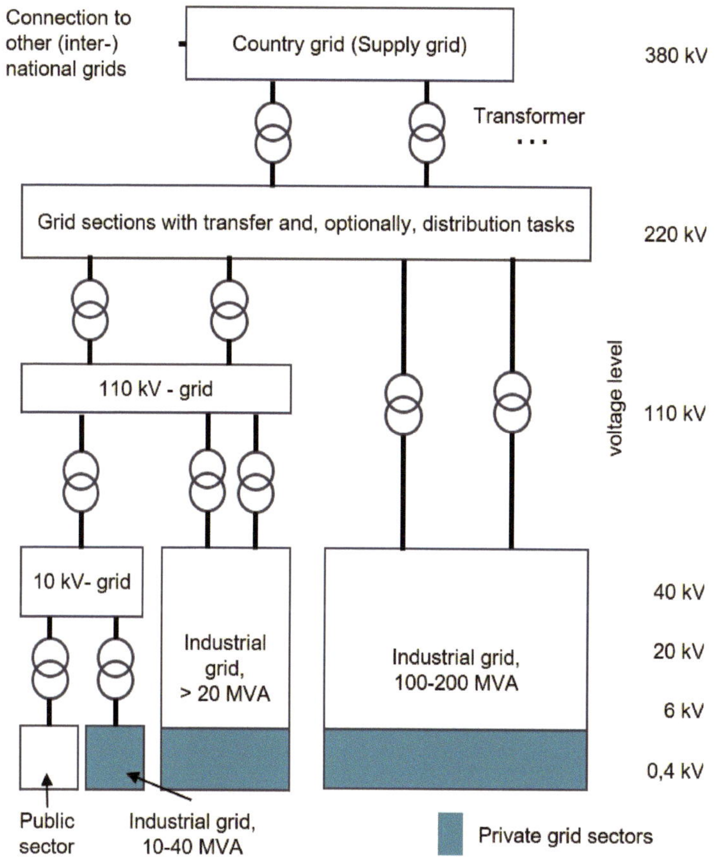


**Figure 8.2** Representation of the structure of the electrical grid [Oedd2004]

In this respect, this chapter is intended to illuminate the planning and design procedure for DC distribution systems from the point of view of a system planner. Technical details of the equipment and the grid, if these are not important for the planner, are deliberately omitted. The aim is to reduce the complexity of implementing a DC grid as much as possible and, in particular, to explore the activities and planning considerations that differ from planning and designing an AC grid.

The project planning of a DC grid includes questions regarding the desired operational management. From the specifications for the operational management of the

DC grid, local energy generation and storage capacities can be derived, whose control by a grid management system must be designed in the further process.

The process of designing a DC grid is based on the planning guide for AC power distribution systems according to Kiank and Fruth [Kian2011]. This process can be adopted as a reference process for the design of a DC grid. The process is supplemented and extended by new design tasks that play no or only a minor role in AC grid structures. Examples of such new tasks are to establish a design procedure for dimensioning the necessary generation and storage capacity within the grid and, on the other hand, the design of a localized grid control, which has so far mainly taken place at higher grid hierarchies.

Figure 8.3 shows the extended planning process. The process begins with the recording of existing grid structures, since planning tasks often aim at expanding existing infrastructure and energy distribution facilities.

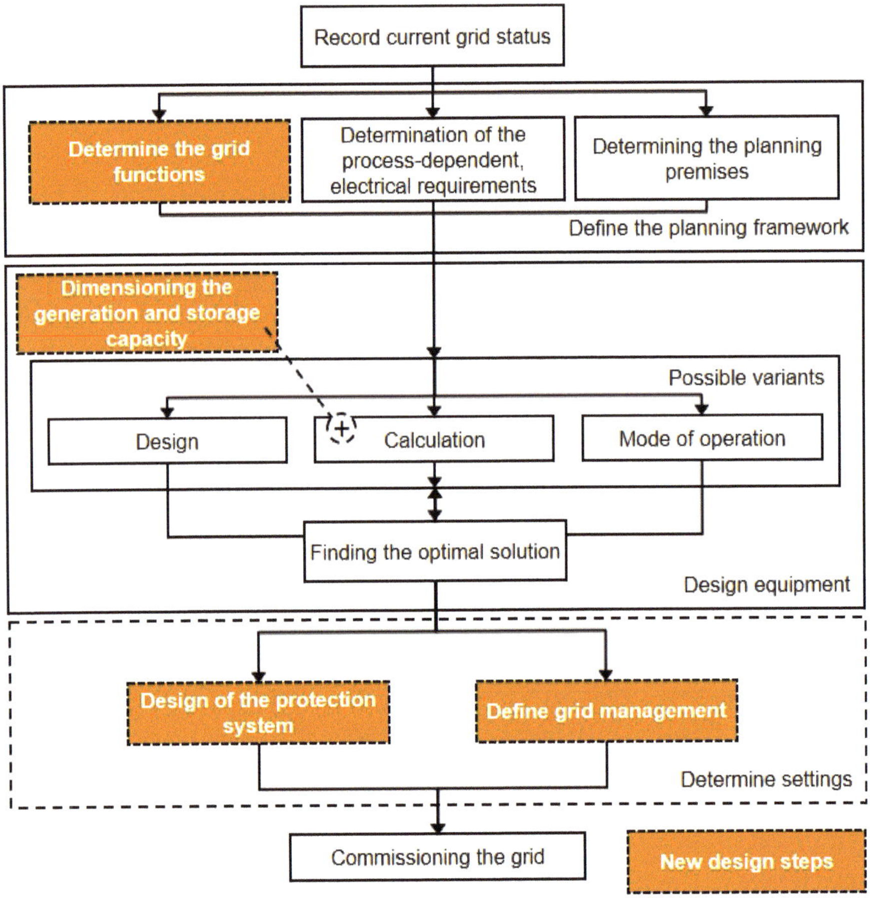


**Figure 8.3** Representation of the extended planning process for energy distribution systems according to [Kian2011, p. 19] with our additions highlighted in orange

Based on this determination of the actual state, the subsequent grid expansion and the necessary planning task are specified in the requirements definition. In the planning of any electrical energy distribution system, the systematic recording of the technical requirements of the consumers with regard to the electrical supply is an essential element. A microgrid structure opens the possibility for the operator to use the grid according to his own ideas, partly independent from the external grid system. This technical design of energy sources, outage protection, prioritization of energy sources and thus self-sufficiency represents a special feature of this planning process.

Several possible solution variants can be compiled from the determined framework conditions. They differ in shape, for example in the grid topology (beam, ring or mesh grid), the rated power of storage systems, the grid feed-in or other dimensioning variables. The operating mode of the grid, such as grounding or the procedure for connecting and switching grid sections, may also vary. This search for the optimal solution can take place regarding economic aspects, such as investment costs, or with respect to technical variables, such as redundancy or availability of the grid. These process steps are summarized as "design of the operating resources".

A major difference in the approach compared to the established design of private grid sections is that a microgrid architecture focuses on the supply of energy from many different sources and allows the intermediate storage of energy. The dimensioning of locally installed storage and generation capacities thus moves into the scope of project planning.

In addition, parameterizations of the devices must be defined in the project planning of a DC power distribution system. On the one hand, these devices include the circuit breakers, which are parameterized to achieve selective protection. On the other hand, the supply power within the microgrid must be balanced, taking into account the instantaneous generation power of renewable sources, the energy stored temporarily, power generation costs in the external grid, and the local energy consumption. The grid management required for this has been described in Chapter 6.

## ■ 8.3 Defining the Planning Framework

At the beginning of any planning, the basic boundary and general conditions must be defined. In addition to the description of the process-related electrical equipment, the required grid functions and the general planning premises have to be determined.

### Determining Process-Dependent Electrical Demands

Each electrical consumer has specific requirements regarding the electrical supply, for example the necessary connection load or the voltage level at the connection point. To take such requirements into account in the design, the first step should be to record the device characteristics. Table 8.1 shows a sample specification for this task. It describes the general properties of the devices that are relevant for operating the latter on the DC grid.

**Table 8.1** Characteristics of Electrical Direct Current Consumers

| General device features | Example values |
|---|---|
| Device designation/device type | DC/DC converter |
| Device class | Class 1 |
| Dimensions (L x W x H) | 100 mm x 100 mm x 100 mm |
| Protection class | IP 20 |
| Rated power/current at the DC connections (at 650 V) | 65 kW/100 A |
| Maximum short-time power/current at the DC connections (at 650 V) | 100 kW/150 A |
| Capacitor technology (electrolyte/foil) | Electrolytic capacitor |
| DC capacity | 2000 µF |
| Longitudinal inductance (e.g., inductance coatings, differential-mode inductance) | None |
| Value of Y capacitors (DC+ to PE) | 47 nF |
| Overload and destruction characteristic ($I^2t$ shutdown limit) | 10 $kA^2s$ |
| DC connection technology | Device connection (screw terminal 10 $mm^2$) |
| Auxiliary voltage 24 V | External supply with 500 mA |
| Fieldbus communication | Profinet with 100 MB |
| Fieldbus connection | RJ45 |
| I/O signals | 6DI/6DO/2AI/2AO |
| Relative static load during precharge | 0.1%/154 mA |
| Discharge function | Passive 5 min |
| DC voltage monitoring | Yes, 10-bit (referred to 900 V) with 1 kHz sampling rate |
| Power monitoring | Yes, 10-bit (referred to max. short-time current) with 1 kHz sampling rate |
| Class(es) for grid management | Extended decentralized group control |

Depending on the type of device, each one has additional, specific properties that have to be recorded, too. These include the characteristics listed in Table 8.2.

**Table 8.2** Specific Device Characteristics

| Device type-specific features | Example values |
|---|---|
| Energy flow direction | Bidirectional |
| Ground fault detection | Residual current monitoring |
| Potential separation | Galvanic separating |
| Rated voltage | 650 V |
| Approved voltage range DC connection | 590 to 800 V |
| Required external protection components | Short circuit protection |
| Max. permissible motor cable length | 5 meters |
| Shielding | Cable shielding required |
| Insulation according to IEC 61010-2-030 | CAT III |
| Safe separation | Yes, according to VDE 0106 T101 |
| Tripping characteristic, tripping time (forward/reverse) | It2 trip characteristic |
| Switching frequency/calculation frequency of digital control loop | 4 kHz/8 kHz |
| Current controller architecture | PI controller with subordinate pulse width modulation |
| Impedance characteristic | Impedance measurement |
| Substitute time constant of the current control | 2 ms |

In addition to the previously mentioned properties, the application produces further data that have to be incorporated into the grid design. For example, the average power consumption of a machine may depend on the product manufactured on it. Table 8.3 lists these characteristics, which mainly concern the power and energy requirements of a consumer. This information will not be known for a new plant, or only in estimates. If similar DC or AC equipment exists, the data can be determined by measurements.

**Table 8.3** Application-Specific Properties

| Application specific features | Example values |
|---|---|
| Dynamics of power change | Boost time until maximum power is reached: 20 ms<br>Load curve: Power as a function of time |
| Average power | 10 kW |
| Maximum power | 500 kW |
| Installation location/plant area | Robot system 2 |

### Determining the Grid Functions

Grid management increases the scope of action of the private grid operator, who can use new types of electrical supply functions. Basically, technical and economic motivations can be distinguished here (Figure 8.4).

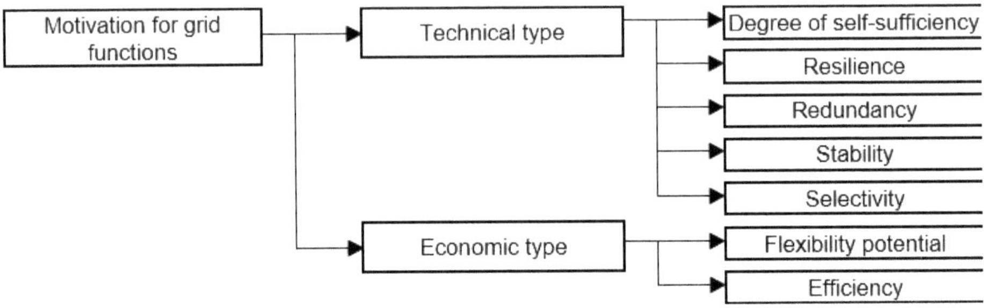


**Figure 8.4** Motivations for grid functions

**Degree of self-sufficiency** refers to the local self-sufficiency regarding electrical energy. This can be obtained with the help of decentralized generators, such as photovoltaic systems, wind turbines, diesel generators or combined heat and power plants. Requirements for optimized operation management can be a specific degree of self-sufficiency, which influences the dimensioning of generation and storage capacities in the grid.

**Resilience**: Conventional power distribution systems rely on a connection to the external distribution grid. Even if distributed generators provide sufficient energy to supply local consumers, these facilities cannot operate as an island grid, or only in exceptional cases. The microgrid inherently includes the ability to bridge the failure of an external supply. Resilience requirements define a bridging period for which the grid will maintain electrical power in the event of failure of either all or only some utilities.

**Redundancy**: The failure of one utility can lead to the failure of the electrical system. If processes and machines are no longer supplied with power, this can lead to production stoppages and damage to products or machinery. If such a supply failure is to have as little impact as possible on the grid, critical infrastructure elements must be provided with redundancy. In the case of a DC microgrid, it should be taken into account that it is fed by different generators. Individual functions can only be fulfilled if the necessary generators are available. Table 8.4 divides the possible realization into different redundancy classes.

Redundancy class one specifies that the grid can always be fed from an unlimited energy source in case of failure of a supply component. If this requirement is met, all storage systems can fail and the microgrid can continue to operate. The disadvantage is that the feed-in power of the generators must be designed for the event that the peak storage systems fail. This leads to a higher connected load, which is actually supposed to be reduced by the use of peak value storage.

Redundancy class two specifies a redundant uninterruptible power supply (UPS) storage system in addition to the controllable generators. Unlike the first redundancy class, the grid does not lose its characteristic of being uninterrupted in the event of a UPS storage failure.

Redundancy class three specifies redundant peak value storages. The advantage of such a design is that the grid can continue to provide short-term load fluctuations from the redundant storage system during the failure of a peak storage device. This has implications for the design of the controllable generators. They do not have to be designed for a complete failure of the peak storage.

Redundancy class four specifies a redundancy of all three energy sources.

**Table 8.4** Redundancy Classes in the Microgrid

| Redundancy class | Coding | Devices to be provided redundantly<br>Controllable generators | UPS storage | Peak value storage |
|---|---|---|---|---|
| 1 | n-1, u, s | 1 | 0 | 0 |
| 2 | n-1, u-1, s | 1 | 1 | 0 |
| 3 | n-1, u, s-1 | 1 | 0 | 1 |
| 4 | n-1, u-1, s-1 | 1 | 1 | 1 |

**Stability** is a fundamental requirement for the grid. The microgrid is decoupled from the external grid and operated with its own control system. This ensures the local balance between feed-in and withdrawal of electrical power. The control of the entire system can become unstable depending on the settings or the selected hardware. Therefore, stability must be checked as a requirement in the design process.

**Selectivity** refers to the selective isolation of a fault (see Chapter 7). It means that only the grid section affected by the fault is disconnected from the grid while the rest of the system can continue to operate. This requirement can apply to all grid sections or only to certain ones.

**Flexibility potential:** "Energy flexibility [is] the ability of a production system to adapt to changes in the energy market quickly and at very low financial cost." [Rein2012] In terms of local electric supply, this characteristic can be understood as a function of grid management. In order to control the power demand from the external grid, the operator of the grid can require that certain consumers are integrated into the grid management or that additional electrical storage capacities are provided.

**Efficiency:** efficiency is – in this context – not understood as pure energy efficiency, but rather as a user-defined objective functional that leads to the selection and prioritization of generators in active grid operation. This functional can be used for economic or non-economic optimization. For example, various cost contributions, such as for carbon dioxide emissions, electricity production costs, or storage costs, lead to minimization of total system costs in operation. Non-economic optimization objectives could be, for example, maximizing renewable generation, maximizing self-consumption, or prioritizing energy sources with low carbon dioxide emissions.

### Determining Planning Assumptions

Planning assumptions are general boundary conditions that determine the framework of the electrical power distribution system.

An important planning premise is the description of the location. This can be included, for example, as a plan of the spatial conditions. Of particular interest are the locations of the planned consumers, as these have a significant influence on the subsequent grid structure and dimensioning. Similarly, the connection point to the external grid or already existing electrical generators, which determine, for example, the location of the grid feed in the DC grid.

The survey is not limited to a two-dimensional representation of the floor, since walls and ceilings also provide space for electrical installations, especially for the planning of lines and conductor rails. Attention should be paid to existing lines that allow for expansion. Furthermore, the location of non-electrical building services must be known in order to identify free installation areas and to avoid hazards and disturbances. The installation of the electrical system in the vicinity of water installations may possibly result in higher requirements for the degree of protection of the equipment against water. Similarly, IT infrastructure is sensitive to coupled electromagnetic radiation.

To ensure that the equipment selected in the subsequent process is used in accordance with the intended use defined by the manufacturer, the climatic environmental conditions of the location must be recorded. In particular, the temperature profile of the environment and the humidity should be taken into account in the selection process. Similarly, the likelihood of a contact with persons of foreign bodies or water entering should be assessed in order to be able to determine the necessary degree of protection for the installations. Guidance on this can be found – for example – in DIN EN 60529 [Deut2014].

Furthermore, the relevant directives and standards that determine partial aspects of a DC distribution system must be observed. A selection is listed in Table 8.5.

**Table 8.5** Selection of Possible Relevant Guidelines and Regulations

| Designation | Relevant aspects |
|---|---|
| DIN EN 60204-1 | Protection against electric shock: discharge times, hazard warnings, disconnecting devices, protection of electrical equipment: motor protection, etc., equipotential bonding, conductors and cables, conductor lines, wiring technology, labeling, warnings |
| DIN EN 60947-1<br>VDE 0660-100 | Low voltage switchgear |
| Generation plants on the medium-voltage grid/AR-N-4105 | Feeding energy from regenerative power generation plants back into the supply grid |
| IEC 60721-3-1/-2/-3 | Classification of environmental conditions |
| IEC 61000-6-4<br>IEC 61000-6-2<br>2014/30/EU | Electromagnetic compatibility (EMC) - Emission requirements for industrial environments |
| EMV-ILA Version 01-04 | EMC Integration Guide for Achieving Electromagnetic Compatibility in Electrical Installations in the Automotive Industry |
| IEC 62477-1 | Safety requirements for power semiconductor converter systems and equipment |
| 2014/35/EU | EU Low Voltage Directive |
| VDE-AR-E-2510 | Stationary electrical energy storage systems intended for connection to the low-voltage grid |
| IEC 62619<br>EN 50272-1/-2 | Safety requirements for lithium accumulators and batteries for use in industrial applications |

## 8.4 Designing the Operating Equipment

The design of the equipment is done in three steps. Firstly, the design and dimensioning as well as the operating mode of the DC grid are defined. Based on this, storage systems are designed and dimensioned. Finally, the selected resources are optimized.

### Design, Dimensioning and Operation

To a large extent, the design of the operating equipment is analogous to the procedure in AC grids. Figure 8.5 shows the analogies as well as the deviating partial aspects of the design. In the following section, the focus is mainly on the deviating design tasks.

### Dimensioning Generation and Storage Capacity

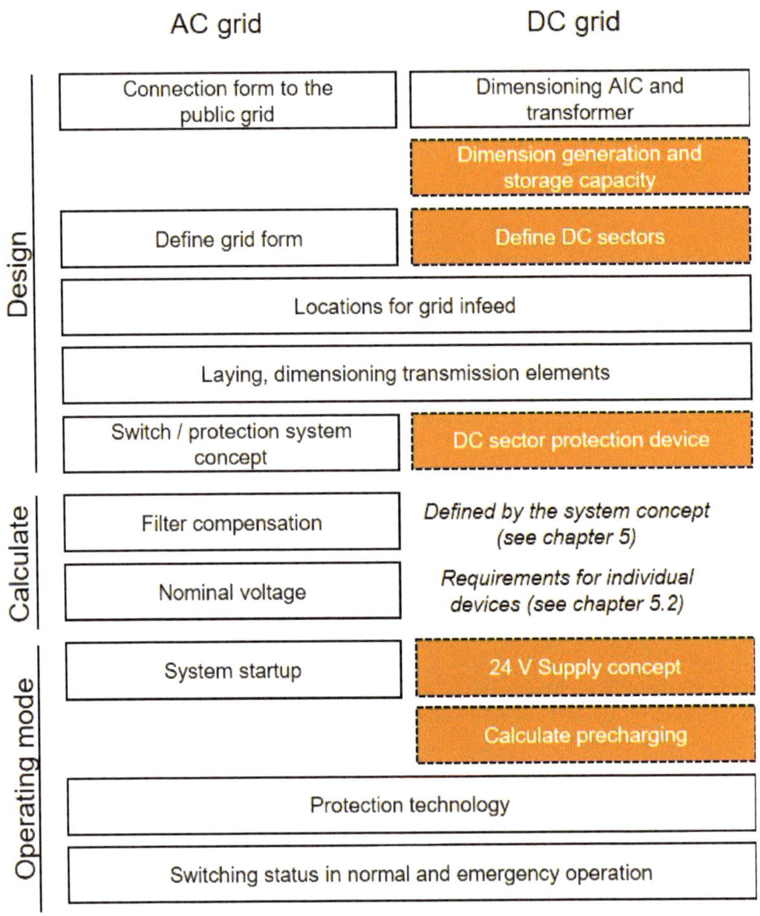


**Figure 8.5** Design, dimensioning and operation tasks

The installation of decentralized generators and storage systems allows the semi-autonomous operation of the DC grid. For the project planner, the planning task is extended by the dimensioning of the decentralized generation and storage capacities in the grid.

The basic motivation for the integration of decentralized generation and storage systems lies both in economic savings in grid connection costs and energy costs and in technical advantages, such as the availability of a supply in the event of a grid failure. The latter secures production and reduces costs incurred by a lack of power supply quality. The sizing of distributed generation capacity is largely based on the expected energy consumption within the supply system. Thus, the sizing of such plants is independent of DC technology. Guidance on the sizing of these plants can be found, for example, in Brauner [Braun2019]. Only the design of storage systems is discussed later in this chapter in detail.

### Define DC Sectors

The DC sector concept has been introduced and defined in Chapter 5.1. As shown there, the division of consumers, generators and storage systems into DC sectors should follow certain rules.

A DC sector is also a protected area. Each DC sector is protected by a Smart Breaker. The grouping of devices in a DC sector should determine – taking into account the device class (cf. Chapter 6.2) – which devices can be disconnected from the grid in one go in the event of a fault within the DC sector.

The DC sectors on the same hierarchical level of the DC grid must be balanced in their average rated power. The reason for this is the requirement for selective disconnection of a grid section in the event of a fault. If there is a short circuit outside the DC sector, reverse currents flow from the DC sector into the higher-level grid section. These return currents flow into the short circuit and subsequently cause recharging currents after disconnection of the fault location. The circuit breakers must be tuned to isolate and disconnect only the DC sector with the fault. This requires that the distribution of capacitances is not too different. Further information on selectivity can be found in Chapter 7.2.

Finally, after the DC sectors have been divided, it should be checked which capacity within the DC sector is already being called up during precharging. For example, sometimes device-internal electronics and simple devices, such as extra-low voltage power supplies, draw power before precharging is complete. This power must be taken into account when dimensioning the precharging circuit (cf. Chapter 5.6).

### Dimensioning Smart Breakers

Smart breakers are used for measuring, switching, precharging, discharging and protection functions (Chapter 5.5) and are to be provided at each junction of the DC grid or before each DC sector. Essential in the sizing of a Smart Breaker is the maximum expected rated current of the DC sector.

The combination of semiconductor and hybrid breakers in one grid is only possible to a limited extent. Thus, only circuit breakers of one technology – hybrid or semiconductor – should follow the fuse protection from a higher hierarchical level. If there are successive fuse cascades in the grid, the circuit breaker technology in the subordinate grid section can be freely selected after a protection with a hybrid switch. In the case of a higher-level protection with a semiconductor switch, a lower-level protection with hybrid switches can lead to problems in the selectivity (see Chapter 7.2).

If the grid configuration requires two Smart Breakers to be cascaded, care must be taken in the case of hybrid breakers to ensure that sufficient capacitance is available between the two.

In grid sections that are subject to regular maintenance, a Smart Breaker with a manually operated two-pole disconnecting device should be provided. After such a disconnecting device, maintenance can be carried out without falling under the scope of DGUV 103-011 – "Arbeiten unter Spannung" [Deut2008].

### Dimensioning the Auxiliary Power Supply

An auxiliary power supply is necessary for operating converters and Smart Breakers. Within the scope of the design, special attention should be given to the fact that an auxiliary power supply must be available in part already before the start-up.

In project planning, it is firstly determined which DC sectors are used for initial precharging of the grid. Auxiliary voltage supply units must be provided in the corresponding DC sectors, which can draw their supply both from the DC grid and from the respective external energy source. For example, the auxiliary voltage supply of a storage DC sector must draw its energy either from the DC grid or directly from the storage system so that it can precharge the grid.

### Defining Precharge

Precharging is conceptually possible without communication between the Smart Breakers. This is possible if suitable switching actions are defined with regard to voltage-time characteristics. These are chosen in such a way that – if the upstream grid section is within the nominal voltage range – a Smart Breaker waits for a certain time before any precharging of a downstream grid section starts. A detailed description of the procedure can be found in Chapter 5.6.

The precharging of a DC sector does not have to be carried out via a precharging resistor (cf. Chapter 5.6). However, if it is provided, then this resistor must be matched in terms of absolute value and power to the associated DC sector parameters.

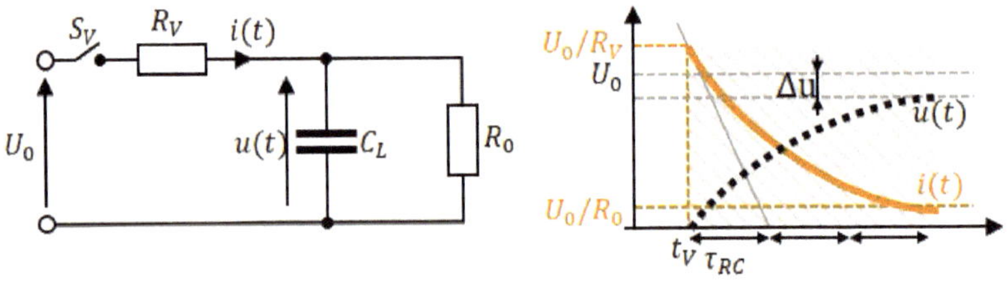


**Figure 8.6** Precharging circuit (left), voltage and current curve (right)

Figure 8.6 shows an equivalent circuit for ohmic precharge. The line voltage $U_0$ is applied in front of the DC sector, which leads to a precharge current via the precharge resistor $R_V$ in the case of the closed switch $S_V$. The DC sector is represented in a simplified way by the installed capacitance $C_L$ and the resistance $R_0$. This resistance represents the power demand $P_0$ already applied during precharging within the DC sector and can be estimated using Equation 8.1.

$$R_0 \approx P_0 / U_0^2 \tag{8.1}$$

The serial connection of resistors $R_V$ and $R_0$ results in a steady-state voltage drop $\Delta u$ across $R_V$ at the end of the precharge. If the maximum voltage deviation $\Delta u$ at the end of the precharge is defined accordingly, a maximum value for the precharge resistance can be specified (Equation 8.2).

$$R_V < \frac{\Delta u}{U_0 - \Delta u} \cdot R_0 \tag{8.2}$$

By specifying a relative charging current $\alpha$ as the ratio of the targeted precharge current $i_V$ and the rated current of the DC sector $i_N$, a minimum magnitude of the precharge resistance can be calculated based on the rated power of the DC sector (Equation 8.3). A suitable resistance value for precharging lies within these limits.

$$R_V > \frac{U_0}{\alpha \cdot P_N} mit\, \alpha = \frac{i_V}{i_N} \tag{8.3}$$

Assuming that the precharge lasts approximately three times the time constant $\tau_{RC}$ (Equation 8.4), an average precharge power $\overline{P}_{RC}$ can be given for this period (Equation 8.5). However, a suitable power value $P_V$ for the resistor is much lower, since the precharge is not a permanent load on the resistor. A suitable overload factor $\beta$ can, according to information from manufacturers, be between 10 and 100 (Equation 8.6).

$$\tau_{RC} = R_V C \tag{8.4}$$

$$\overline{P}_{RC} = \frac{C U_0^2}{2 \cdot 3 \tau_{RC}} + P_0 \tag{8.5}$$

$$P_V \approx \frac{\overline{P}_{RC}}{\beta} \tag{8.6}$$

### Design and Sizing of Storage Systems

Energy storage systems are an important key component for the DC grid and offer various uses, especially in increasing the security of supply, reducing energy consumption (use of recuperation energy), reducing peak loads, and providing protection during grid outages. The procedure for selecting and dimensioning the energy storage system is largely dependent on the selected use case. In addition to the application, the subsequent load profile in the DC grid should be known. The load profile refers to the power consumption of all consumers as a function of time. This knowledge is necessary in order to draw appropriate conclusions about the amount of energy required.

*Possibility of using an energy consumption optimization*

Energy consumption optimization means the use of recuperation energy. The latter can be generated, for example, by the braking process of a machine axle in a production process. The kinetic energy is converted into electrical energy and fed back into the DC grid. This excess electrical energy would have to be either converted into heat via braking resistors or fed back into the AC grid in the event of a total power surplus in the DC grid.

With energy storage in the DC grid, however, recuperation energy can be stored temporarily. If a load peak occurs later, the temporarily stored energy can be retrieved. With a known load profile, an initial design of the energy flows can be calculated. Figure 8.7 shows the effect on the resulting load profile.

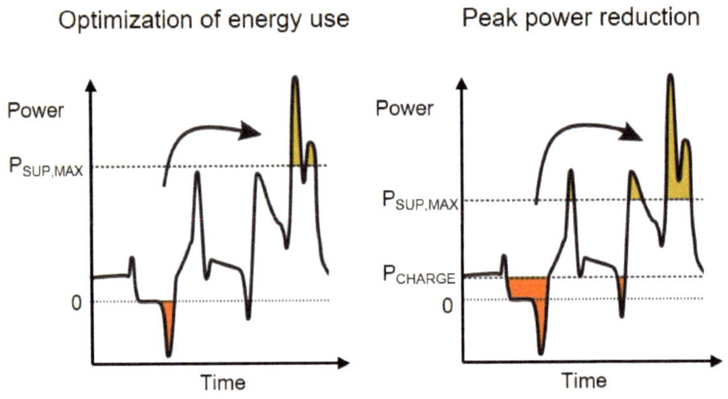


**Figure 8.7** Possible uses – energy consumption optimization (left) and peak load reduction (right)

The recuperated energy (orange) is absorbed via the energy storage system and supplied to the process later. In the next step, the power limit $P_{SUP,MAX}$ is determined, above which the energy storage system supports the supply of energy (yellow area). It should be noted that there must be a permanent balance of recuperated energy and returned energy above the power limit $P_{SUP,MAX}$. Otherwise, the energy storage would discharge or charge to its maximum capacity. In both cases, economic potentials would be lost. Figure 8.8 shows the load profile of an industrial robot after energy consumption optimization.

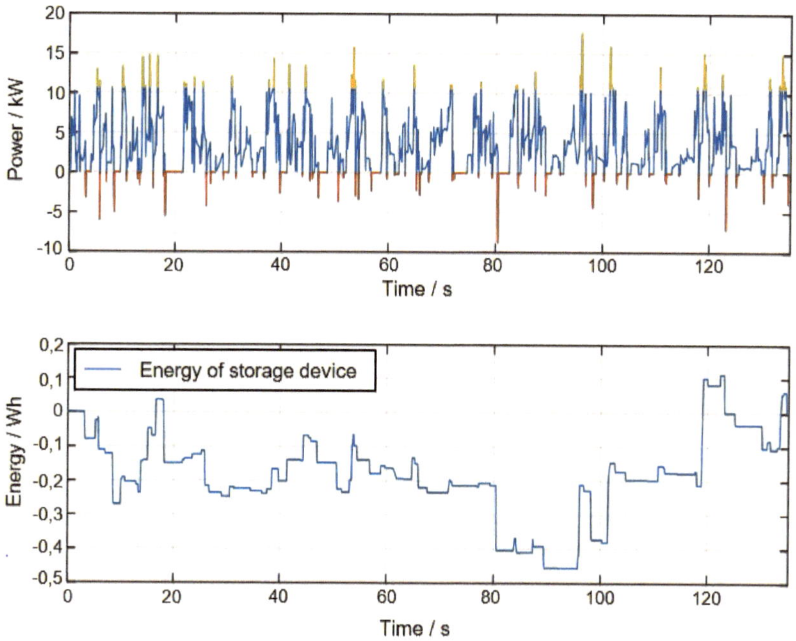


**Figure 8.8** Analysis of a load profile after energy consumption optimization (top) and course of energy

*Possibility of using a load peak reduction*

Load peak reduction, also called load peak shaving, enables a reduction of the grid connection power. Load peaks that occur are not served exclusively by the AC grid supply, but in part by the energy storage system. This also leads to a stabilization of the energy supply from the AC grid.

One possible approach is for the plant operator to firstly specify the maximum power for the AC grid connection $P_{SUP,MAX}$. In this case, the energy storage system is designed so that if the power demand in the DC grid is higher than $P_{SUP,MAX}$, the energy storage system supports the grid and is itself discharged in the process. Design variables in this case are the power and capacity of the storage. To calculate the maximum discharge power of the energy storage device $P_{ESS,MAX}$, the maximum load demand in the load profile max*(P)* must be determined (Equation 8.7).

$$P_{ESS,MAX} = \max(P) - P_{SUP,MAX} \tag{8.7}$$

By numerical integration of all power points above $P_{SUP,MAX}$ the required capacity is obtained. In the right part of Figure 8.7 the yellow area represents this area integral. Subsequently, the power $P_{CHARGE}$ should be determined, from which the energy storage must be recharged, so that it is again in the same state of charge after running through a load profile. The power is also calculated by numerical integration of the load profile (see Figure 8.7 right - orange area). The charging energy (orange area) and discharging energy (yellow area) must be equal in area, taking into account the charging efficiencies within a load profile. In this way, equilibrium of the state of charge is achieved.

*Possibility of use load peak reduction for safeguarding against power failures*

Energy storage units in the DC grid also offer a simple way of implementing an uninterruptible power supply (UPS) and thus safeguarding production against grid failures. This was tested, for example, on a CNC machining center (see Chapter 6.4 and Chapter 9.1). Two points are decisive for the design of an energy storage system for an uninterruptible operation of the DC grid. First, the maximum connected load: If one or more energy storage units are to temporarily take over the complete power supply, they must be able to meet the power requirements of all loads, i.e. the load profile of all loads in the DC grid must be able to be served by the energy storage units. Secondly, the system operator must define the maximum time period in which the energy storage takes over the complete power supply. The load profile can then be used to calculate the amount of energy to be held in reserve.

**Finding the Optimal Solution**

The design of the DC grid described in the previous section does not lead to an unambiguous solution. Rather, the project planner has a large number of degrees of freedom in the design in the course of the complex design task. It is part of the project planner's task to check the solutions created against the planning framework and, if necessary, to define further evaluation criteria. For example, the costs of the system can be a target-oriented evaluation standard to decide between different realization options in this phase of the design.

Regarding the planning of DC grids this procedure does not differ from the procedure taken regarding AC grids. Further literature is found in [Kian2011].

## ■ 8.5 Determining the Settings of the Transformer Systems and Protection Devices

Even more than in AC grids, devices in the DC grid can be configured by adapting digital regulation and control systems. This chapter directs the focus to the adjustment of the converter systems that connect the DC grid with the external AC grid, regenerative energy sources or storage systems. Information on dimensioning and determining the settings of the protection technology can be found in Chapter 7.2, where the challenges of protection technology in the DC grid are elaborated upon and important design aids for project planning are given.

The concept of the DC factory stipulates the grid control options described in Chapter 6.2. Firstly, the decentralized grid control with control characteristics allows to balance the power of all currents in the grid. On the other hand, this simple control based on the decentralized voltage measurement can be used to realize desired grid functions of the operator without using an additional energy management system. The literature calls this approach "Digital Bus Signaling". It has been transferred to the application for industrial applications [Schö2006] [Gavr2014] [Xiao2013] [Hwan2015].

Figure 8.9 shows a process model and the tasks for determining appropriate settings for the DC grid. Input data into the process are the planned *grid functions* from the requirement definition. In addition, there are the device data of the loads and feed-in devices used (*device description*), as well as their arrangement in the grid and the physical properties of the grid control.

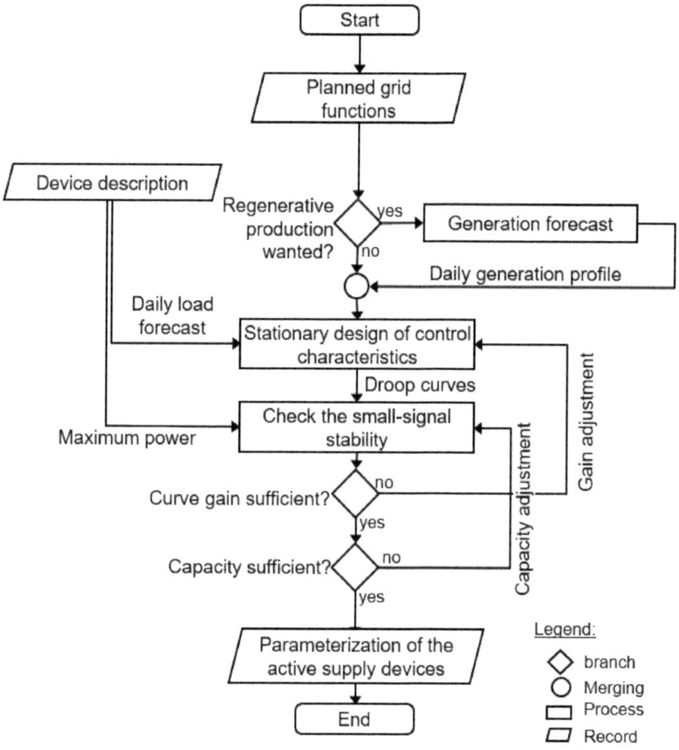


**Figure 8.9** Procedure model for the design of a grid control

The design procedure assumes that a daily forecast of the renewable generation is available or can be estimated. This is included in the steady-state design of the control characteristics together with the daily load forecast.

The overall gain (slope of the control characteristics) over the voltage range as well as the maximum expected load influence whether the system is small-signal stable (cf. Chapter 6.3, [Ott2019]). If the overall gain is not sufficient, it must be increased to a suitable value in an initial step. If the requirement for small-signal stability is jeopardized by this adjustment, additional capacity may have to be provided in individual DC sectors.

### Steady-State Design of Control Characteristics

With characteristic-based control, the grid voltage can be used as an information carrier about the state of the grid (Section 6.4). This means that the voltage is used to selectively activate or deactivate energy sources and consumers[1] in the grid in certain situations.

Therefore, the first step is to identify stationary target states of the grid from the user-defined requirements. This is shown in Table 8.6 as an example for a grid with renewable energy sources, energy storage systems and connection to an external distribution grid. In this case, the user's requirements regarding the functional fulfillment of the grid management within the DC grid are a steadying of the grid power supply, the maximization of self-consumption of regeneratively generated energy, and a robust system behavior in case of a fluctuating regenerative feed-in or consumption situation.

Based on the identified target states, stress bands are to be defined and assigned to the states (Table 8.7). Subsequently, the voltage bands are to be arranged in a logical order. Starting from an average voltage value, this arrangement mostly follows the logic of an increasing energy surplus with increasing voltage as well as of an increasing use of more expensive forms of energy with decreasing voltage.

[1] Consumers can be integrated into grid control if they can adjust their output within certain limits independently of external requirements. This can be used, for example, to lower consumption in cases of low renewable generation.

**Table 8.6** Linking the Stress Bands to the Steady-State Target Conditions

<table>
<tr><th>No.</th><th colspan="3">Stationary target state</th><th>Naming</th><th>Voltage limits</th></tr>
<tr><td colspan="2">Grid feed</td><td>Storage system</td><td>Regenerative generation</td><td></td><td></td></tr>
<tr><td>1</td><td>Grid feeders supply a constant maximum current to the grid</td><td>Storage systems support the grid by increasing the discharge current linearly with decreasing voltage</td><td rowspan="4">Renewable sources feed maximum available power into the grid</td><td>Peak load capping</td><td>$U_{UP} - U_{PS}$</td></tr>
<tr><td>2</td><td>Grid feeders adjust their feed-in current linearly around the average current draw of the grid</td><td>Storage systems do not provide electricity</td><td>Reduced grid feed</td><td>$U_{PS} - U_{IC}$</td></tr>
<tr><td>3</td><td>Grid feeders provide a constant minimum current to the grid</td><td>Charging current increases linearly with increasing voltage</td><td>Charging energy storage systems</td><td>$U_{IC} - U_{GR}$</td></tr>
<tr><td>4</td><td>Grid feeders reduce their feed-in current below the minimum limit and start feeding back into the AC grid</td><td rowspan="2">Storage systems charge with constant maximum current until the charging limit is reached</td><td>Mains regeneration</td><td>$U_{GR} - U_{GS}$</td></tr>
<tr><td>5</td><td>Grid feeders provide maximum available regenerative current to the AC grid</td><td>Feed-in current of renewable sources is linearly reduced with increasing voltage to prevent overloading of the grid</td><td>Generation stop</td><td>$U_{GS} - U_{ES}$</td></tr>
</table>

Figure 8.10 shows the arrangement of the voltage ranges in relation to each other with reference to the previously considered example. The steady-state voltage ranges are to be arranged within the nominal operating range defined in the system concept (cf. Chapter 5.3). The characteristic curve design first determines a sum characteristic, which is broken down to individual devices in the next step. In this process, all devices present in the grid must be assigned to a participant class of the grid management. This is shown as an example in Table 8.7.

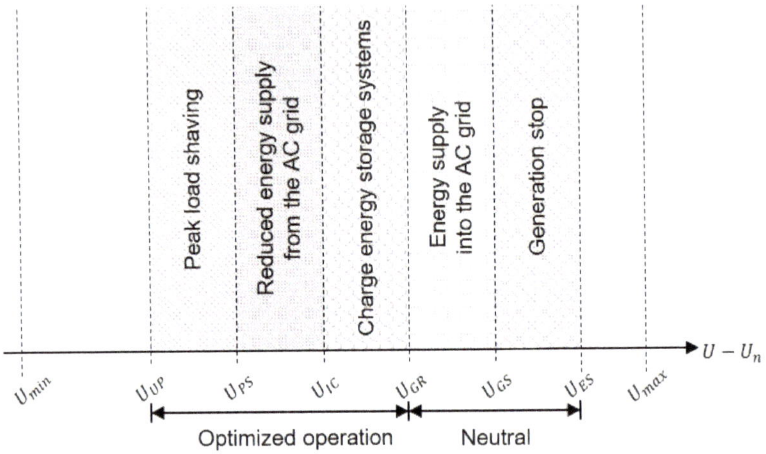


**Figure 8.10** Functional voltage ranges in the DC grid

For the definition of the characteristic curves, the respective voltage limits represent a boundary condition for the current requested by the participant class. These can be maximum modulation limits, such as the upper voltage limit $U_{ES}$. Such a control limit is characterized, for example, by the fact that the participant class "Regenerative Generation" sets its feed-in at this point to zero and the participant classes "Grid Feed-in" and "Energy Storage" have a maximum return or feed-in value (Figure 8.11).

**Table 8.7** Exemplary Assignment of Real Devices to Participant Classes of Grid Management

| Participant class | Assigned, real devices | Maximum feed current | Maximum regenerative current |
|---|---|---|---|
| Grid feed | 1. active infeed converter 1 (rated power 200 kW) | -800 A | 800 A |
| | 2. active infeed converter 2 (rated power 350 kW) | | |
| Energy storage systems | 1. lithium storage system (capacity 1 MWh, connection with DC/DC converter, nominal power 200 kW) | -300 A | 300 A |
| Regenerative generation | 1. photovoltaic system (peak generation capacity 200 kW, connection via DC/DC converter) | -1050 A | 0 A |
| | 2. wind turbine (peak generation power 500 kW, connection via DC/DC converter) | | |

The second type of voltage limits are sharing limits (range transitions), which are used to take additional power requests from the grid by another participant class above a certain current or power limit value. An example of this is $U_{PS}$, which determines from which feed-in power the grid feed-in is supported by the energy storage systems.

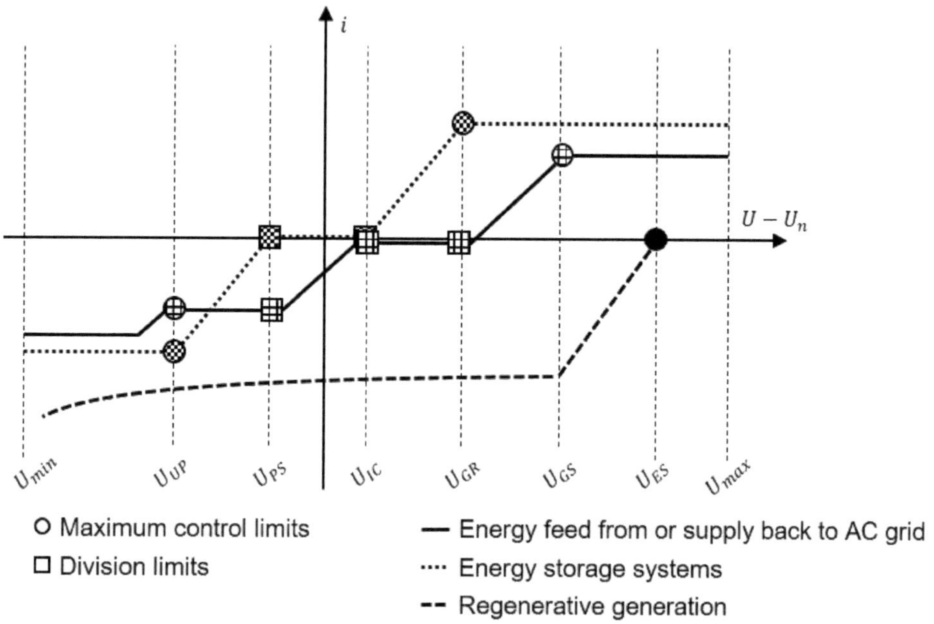


**Figure 8.11** Representation of the limits and the resulting characteristic curves

### Checking the Small Signal Stability

The small-signal stability of the grid is related to the characteristic slope in the particular voltage section. On the one hand, there is the requirement for a quasi-stationary design of the characteristic curves, which means operating within the desired and permitted voltage band. In principle, the steeper the characteristic curves are designed, the smaller the voltage range can be selected. This is opposed by the decreasing damping of the systems when the slope is increased (see Chapter 6.3).

For the design of the overall grid, the capacity available in the grid limits the maximum permissible slope of the characteristic curve. Therefore, it makes sense to first determine the desired characteristic configuration and then assign the necessary capacity to each DC sector according to Chapter 6.3 Equation 6.3. If the capacitance of a DC sector is not sufficient, the designer must ensure that the device does not violate the required quality of the output impedance. In addition, it must be taken into account that in the event of a fault not all grid capacitances are switched on. A sufficiently large tolerance must be taken into account for the quality of the output impedance.

If electrical loads are operated with a power electronic converter, these systems can generate instabilities. To avoid this, the impedance of the grid should always be below the impedance of the load.

From a design perspective, this danger can be minimized. For this purpose, in a first step, the resulting characteristic slope $G_{\mathrm{m,res}}$ of all n devices must be determined for each voltage band $m$ (Equation 8.8).

$$\frac{1}{G_{m,res}} = \sum_{i=1}^{n} \frac{1}{G_{i,m}} \tag{8.8}$$

Subsequently, the minimum quasi-stationary internal resistance of the total load $R_{\mathrm{m,load}}$ is determined using Equation 8.9. This takes into account the respective minimum voltage $U_{\mathrm{m,min}}$ of the voltage band $m$ as well as the maximum power consumption $P_{\mathrm{max}}$ to be expected at the grid.

$$R_{m,Last} = \frac{U_{m,min}^2}{P_{max}} \tag{8.9}$$

To avoid instabilities, for each voltage band m the reciprocal of the resulting characteristic slope $G_{\mathrm{m,res}}$ should always be smaller than the quasi-stationary internal resistance of the total load (Equation 8.10).

$$\frac{1}{G_{m,res}} < R_{m,Last} \tag{8.10}$$

Further information on the stability of the control system can be found in Chapter 6.4.

## Literature

[Kian2011] Kiank, Hartmut; Fruth, Wolfgang: Planungsleitfaden für Energieverteilungsanlagen. Konzeption, Umsetzung und Betrieb von Industrienetzen. Publicis Publishing, Erkangen, 2011

[Scha2019] Schaab, Darian Andreas.; Bianchini, Isabella; Sauer, Alexander: Design Framework for the Grid Control of a Multiple Infeed Industrial Direct Current Microgrid. 2019

[Rein2012] Reinhart, G; Reinhardt, D; Graßl, M; Einführungen zur Bewertung der Energieeffizienz von Produktionssystemen. wt Werkstattstechnik online 102 (2012), Nr. 9, pp.622-628, Springer-VDI-Verlag, Düsseldorf, 2012

[Deut2014] Deutsches Institut für Normung: Schutzarten durch Gehäuse (IP-Code) (IEC 60529:1989 + A1:1999 + A2:2013) Deutsche Fassung EN 60529:1991 + A1:2000 + A2:2013,Beuth Verlag GmbH, Berlin, 2014

[Braun2019] Brauner, Günther: Systemeffizienz Bei Regenerativer Stromerzeugung. Strategien Für Effiziente Energieversorgung Bis 2050. Springer Vieweg, Wiesbaden, 2019

[Deut2008] Deutsche Gesetzliche Unfallversicherung: DGUV Regel 103-101:2008, Arbeiten unter Spannung an elektrischen Anlagen und Betriebsmitteln. Carl Heymanns Verlag, Köln, 2008

[Schö2006] Schönberger, John; Duke, Richard; Round Simon: DC-Bus Signaling: A Distributed Control Strategy for a Hybrid Renewable Nanogrid. IEEE Transactions on Industrial Electronics, IEEE, Piscataway Township, 2006

[Gavr2014] Gavriluta, Catalin; Candela, Jose Ignacio; Citro, Costantino; Rocabert, Joan; Luna, Alvaro; Rodriguez, Pedro: Decentralized Primary Control of MTDC grids With Energy Storage and Distributed Generation. IEEE Transactions on Industry Applications, IEEE, Piscataway Township, 2014

[Xiao2013] Xiao, Jianfang; Peng, Wang: Multiple modes control of household DC microgrid with integration of various renewable energy sources. IECON 2013 39th annual conference of the IEEE Industrial Electronics Society, IEEE, Vienna, Austria, 2013

[Hwan2015] Hwang, Pyeong Ik; Jang, Gilsoo; Pyo, Gi Chan; Han; Byung Moon; Moon, Seung Il; Ahn, Seon Ju: Dc microgrid operational method for enhanced service reliability using DC bus signaling. Journal of Electrical Engineering and Technology, Korea, 2015

[Ott2019] Ott, Leopold: Modellbasierte Stabilitätsuntersuchungen in Niederspannungs-DC-Verteilnetzen. Dr. Hut - VerlagMünchen, 2019

[Oedd2004] Oedding, Oswald: Elektrische Kraftwerke und Netze. 6. Auflage, Springer-Verlag, Berlin, Heidelberg, New York, 2004

[Deut2005] Deutscher Bundestag: Gesetz über die Elektrizitäts- und Gasversorgung/Energiewirtschaftsgesetz. EnWG, https://www.gesetze-im-internet.de/enwg_2005/BJNR197010005.html#BJNR197010005BJNG001000000 (access on 21 Dec 2018)

# 9 Implementation Examples

Timm Kuhlmann
Herbert Graf
Patrick Kröhner
Tobias Laubscher

Within the DC-INDUSTRIE research project, the numerous developments - from the system concept to safety devices to grid management - were tested in four different plants. In each, the objectives of testing the DC technology differed. In particular, the type of supply and the characteristics of the grid management were varied and adapted to the application context. In the following, the application cases are presented in detail, the respective objectives are described, and the main results summarized. The DC grid as described in this book worked in all plants. Therefore, what could be proven is: The system concept and grid management are effective. The expected potentials can be tapped. The requirements are met.

## 9.1 CNC Processing Center

The first application example is a CNC processing center from HOMAG - similar to the one in Figure 9.1 - for processing panel-shaped workpieces made of wood or wood-like materials. The machine has eight controlled axes that freely position the processing portal above the surface, tilt the drilling and milling spindles and plunge them into the workpiece. The processing portal moves above the workpiece and mills the contour or drills the holes into the workpiece on the machining table. The power of the drives in the machine varies between 15 kW for the milling spindle and 0.9 kW for smaller axes and tool changer. In total, there are 15 drives in the machine, each of which is operated by a separate converter. The total power drawn is up to 22 kW during machining.

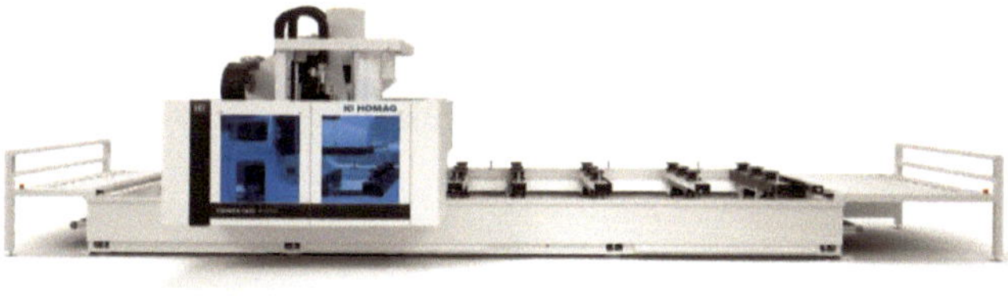


**Figure 9.1**
HOMAG CNC processing center

Regarding the local concentration of the drives, this is a small energy supply grid. Therefore, a DC supply was tested that is simple in terms of its internal structure and thus cost-effective to implement. A 50 kW B6 bridge rectifier was selected. This does not allow targeted grid management; however, it was not required with respect to this machine. Nevertheless, the aim was to reduce the peak power at the connection point of the machine by means of an energy storage device for power peak limitation. Depending on the energy situation, this storage had to partially take over the supply of up to 10 kW or recharge itself during off-peak periods. The high demands on positioning accuracy during movement made it necessary to control all axes separately using a control unit developed at HOMAG. This means that even complicated contours can be milled with very high precision. These motors, which had previously been operated separately in terms of energy, were energetically connected to the DC grid so that an energy exchange could take place. The entire machine was divided into four DC sectors (Figure 9.2).

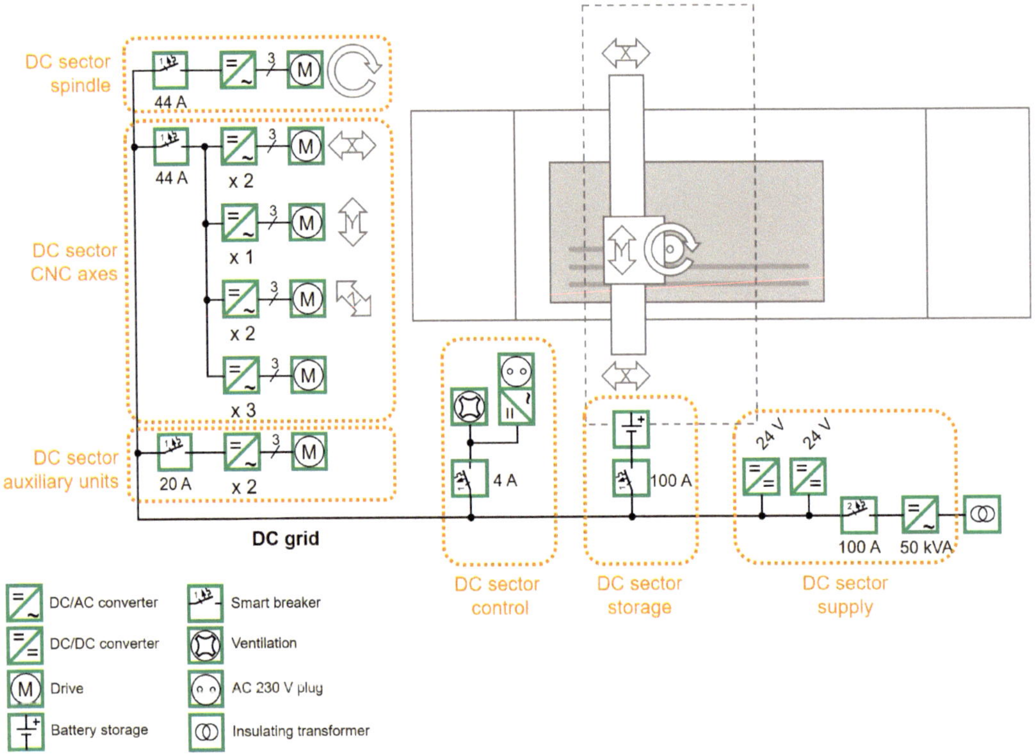


**Figure 9.2** Structure of the DC grid with all connected participants of the CNC machining center

The DC sector "spindle" contains the drives for the milling spindle and the drilling gear. All drives for positioning and tool change are combined in the "CNC axes" DC sector. The auxiliary units for the vacuum pump and chip transport are located in the "auxiliary units" DC sector. All control components were grouped in the "control" DC sector. For the supply of the machine, a feed-in DC sector was formed for the mains connection and another feed-in DC sector for the energy storage. The DC sectors are each protected by a Smart Breaker. The Smart Breaker for the feed-in from the AC grid has two poles, since no isolating transformer was connected upstream of the Smart Breaker. The grid is therefore a TN-grid.

During the operation of the machine with DC supply, no change in the mechanical behavior was observed. The machining of the workpieces was performed in the same way with the same speed and precision. Regarding the energetic load profile, a significant change could be identified.

Figure 9.3 shows that in the DC grid, the braking milling spindle feeds its energy back into the DC grid.

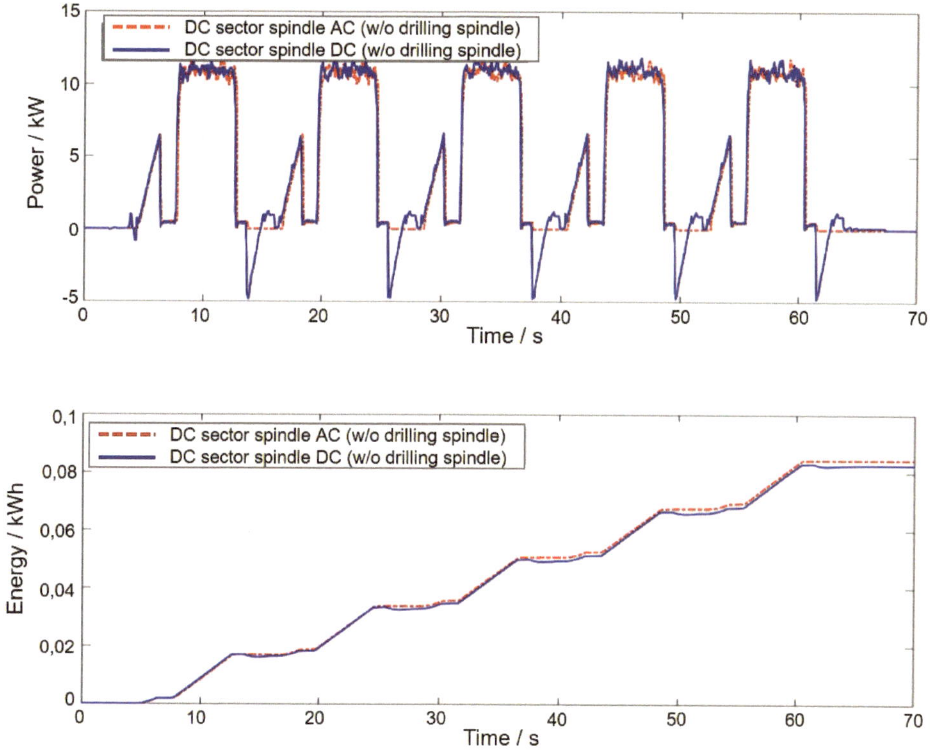


**Figure 9.3** Recovery of power (top) and energy (bottom) during braking of the milling spindle

Other drives that have to be supplied with energy at the same time absorb the braking energy. In this example, the power of the drilling spindle was also measured when measuring the DC system. It can be seen that the total energy consumed is even slightly lower than is the case with the AC system, despite the additional consumer at the measuring point. Therefore, the additional consumer can therefore be supplied directly by the recovered energy of the decelerating milling spindle.

This reduces the total energy demand of the machine for one machining cycle. Figure 9.4 shows the measured power in each case at the main infeed of the original AC and the converted DC system.

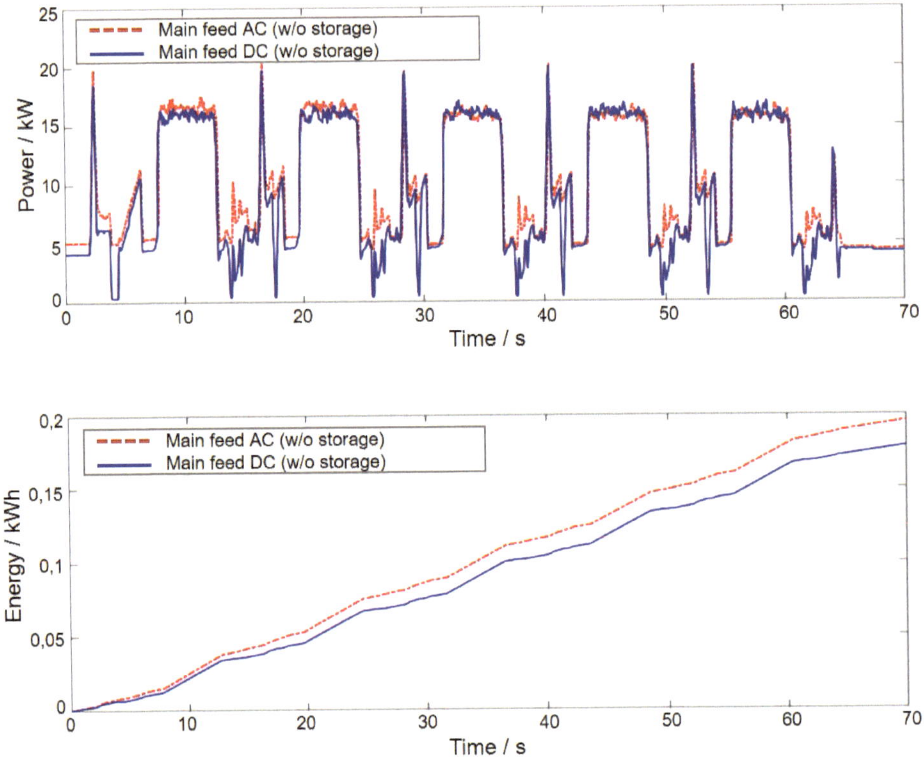


**Figure 9.4** Power comparison (top) and energy comparison (bottom) AC/DC at the main feed

Regarding the reference workpiece used for validation, the saving amounts to 0.513 kWh per processing hour, which in turn means an energy saving of approx. 5.5%. The brake choppers are not loaded and remain in the system only for emergencies, i.e., if no energy absorption by other participants is possible.

Peak load reduction is achieved by using the accumulator. It has a control characteristic that supports the supply of the system as soon as a certain power demand is exceeded. As the power demand at the utility increases, the voltage in the system decreases. The latter is due to the higher voltage drop across the internal resistance of the utility during the period in which the current increases. This voltage drop is registered by the DC/DC converter of the storage system. The battery supports the system. If the power consumption drops, the voltage rises again. The storage system reduces the supply or can recharge from the mains when the machine's power requirements are low. The power consumption and thus also the energy consumption of the machine at the connection point are stabilized. Figure 9.5 shows the power consumption of the machine with the storage system and that of the storage system itself.

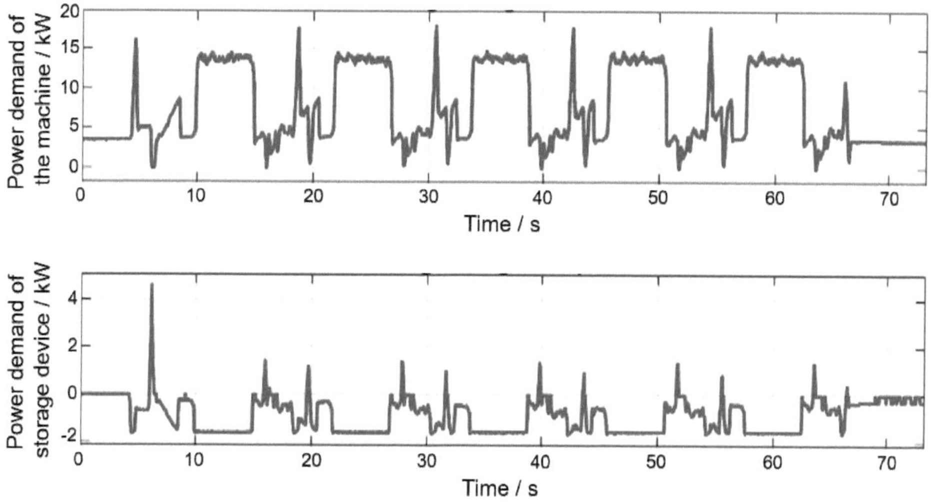


**Figure 9.5** Power consumption of the machine (top) and the accumulator (bottom)

The limit value was set to 15 kW in the test scenario. This reduced the maximum power consumption of the machine from 22 kW without storage to 15 kW with storage. A further reduction down to the average power is possible, but it significantly influences the necessary storage capacity and the performance of both the storage and the DC/DC converter. It thus drives up costs. An economic balance must be found between the storage system costs and the savings from load reduction and reliability.

In terms of the DC grid within the machine, the cost-saving potentials lie in converters without an AC supply unit, in lower dimensioning of brake choppers and in filters. While DC protection technology is still significantly more expensive today – but also more functional than conventional AC technology – it can be assumed that prices will fall as the market develops. As soon as series devices from different suppliers have established themselves on the market, cost parity for the electrotechnical drive and supply components in the AC and DC supply grid can be expected considering a network of eight machine tools.

## ■ 9.2 Intralogistics Systems

Another application example for an open DC grid in a production plant was "Intralogistics System with Handling Stations and Roller Conveyors". The test plant consists of two handling stations connected by roller conveyors. The first handling system is a palletizing system which stacks beverage crates onto a Euro pallet. This loaded Euro pallet is then transported by roller conveyors to the second handling system, a depalletizer. This part of the line separates the crates from the Euro pallet again. The empty crates and the empty pallets are then transported back to the palletizing system via separate conveyors. All in all, this creates a closed-loop system that has real plant sections connected in a confined space and can be operated in an endless loop. The entire system uses 30 drives from different manufacturers in a power range between 0.75 kW and 22 kW. In the AC case, the drives for the axes of the palletizers are each operated in a servo axis network with energy exchange via the internal DC link. Energy storage is not used in this application example. The objective is not to reduce peak loads as in the first application example, but to exchange energy between the numerous drives from different manufacturers within the system. In addition, the aim was to analyze the technical feasibility of a decentralized drive control with converters directly on the motor and thus longer DC cables. Power is supplied via a central Active Infeed Converter (AIC). Figure 9.6 shows the structure of the plant with the palletizers and the roller conveyor systems.

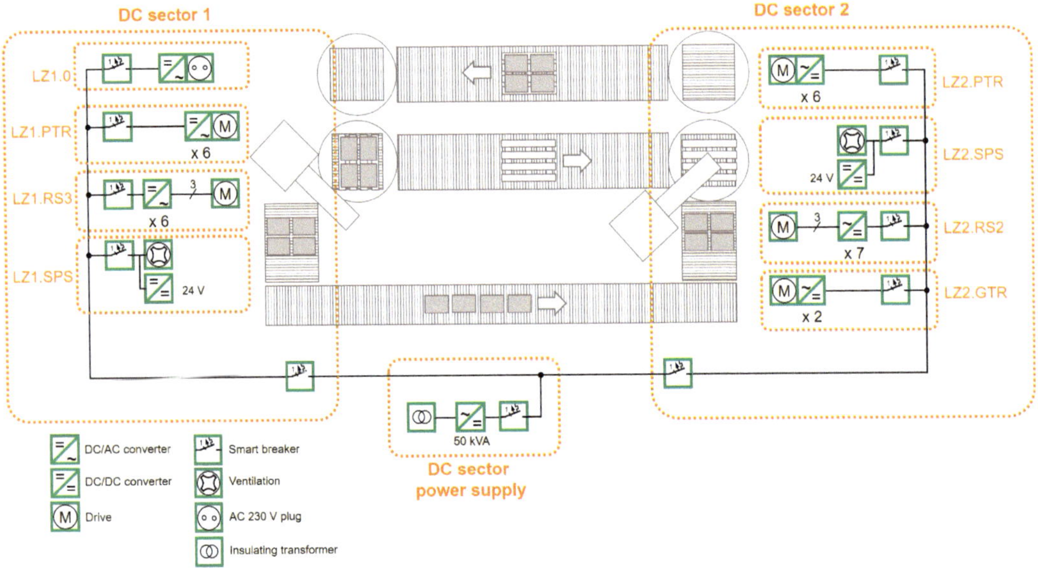


**Figure 9.6** Plant overview of the application example with handling systems and roller conveyors

Similar to the first application example, i.e., the machine tool, this system was first set up and commissioned using conventional AC technology. Subsequently, measurements of the power consumption and the energy demand as well as the current characteristics in subsystems were measured as reference data. Only then was the system converted to DC technology. In the process, the drives of the roller conveyors were equipped with decentralized converters.

The plant is divided into three DC sectors. DC sector 1 (left side in the picture) contains the drives for the depalletizer as well as the adjacent conveyor technology and the necessary air conditioning technology in the control cabinet as well as the control system itself. DC sector 2 (right side in the picture) contains the components of the palletizer. The supply by means of the AIC with isolating transformer forms another DC sector. The powers of DC sectors 1 and 2 are of a similar order of magnitude with a deviation of about 20%. Within the DC sectors, subordinate DC sectors were formed. These group the drive systems according to their plant affiliation. In each case, the conveyor drives and the drives for the handling systems were grouped together. In addition, DC sectors were formed for an AC 230 V service socket, which is supplied from the DC grid, and for the supply of the air-conditioning units and the controls.

The application example shows that a DC grid in a more extensive plant with components from different manufacturers works without any problems. The decentralized motor converters with DC supply could be built more compactly due to the adapted cooling concept. This means that decentralized components can be used more efficiently. The energy exchange between the drives was extended beyond the existing servo axis network. The energy exchange between the two machines now took place within the DC grid, thus avoiding conversion losses when feeding back into the AC grid. The energy savings achieved in the plant were a sum of DC efficiency potentials and higher efficiency by decentralizing the converters. The EMC limits as given in Chapter 5.7 were complied with. Subsequent shielding of the DC lines was not necessary, thereby fulfilling the important requirement of not having to shield any additional lines.

The costs for the components of the drive system without the motor and the DC supply are estimated to be slightly higher for the DC test application considered than would be the case regarding the system costs of the AC application. The cost saving potential lies in the centralization of the DC supply for the drives and thus in a lower dimensioning of the peak power. Again, the cost of switching and protection drives the overall cost of DC realization. For larger systems - roughly four times the systems in the test application - cost parity between DC and AC can be expected. This is due to the further reduction in the peak power of the supply device and the associated relative reduction in the cost of protection technology.

**Figure 9.7**
Electric monorail system in the Tec factory of Daimler AG

Another application example for the realization of a DC grid in an industrial environment is an electric monorail system test facility (EMS), see Figure 9.7. Five trolleys, which can raise and lower the car body, travel on a rail ring divided into five segments. The trolleys are supplied with DC voltage via the rail. The connection is made via sliding contacts. The drive controls for the individual carriages and their lifting drives are positioned at the top of the carriage.

Power is supplied to the rail separately for each segment with a feed line from the DC supply. The traction drive is a 0.75 kW motor, the lifting motion is enabled by means of two 7.5 kW drives. Due to the gears and the mechanics of the lifting device, recuperation of energy during lowering is not possible. Figure 9.8 shows the structure of the DC grid for the EMS. What is special at this point is that the DC sectors with the connected participants do not remain constant. The movement of the cars causes crossovers from one DC sector to the neighboring one, so that empty DC sectors and DC sectors with up to three cars also occur. While the protection of the individual trolley remains unaffected by this, the maximum number of trolleys is an important dimensioning parameter for the protection of a single rail segment in the distribution.

During operation of the test application, it was shown that a change of the participants and loads within a DC sector had no effect on the system. Even a brief connection of two DC sectors via the contact of the trolley did not cause any disturbances in the DC grid. This means that a DC grid can also be used for dynamic grid structures that change during operation. With regard to the load profile, there were no significant differences between the AC and DC realizations. Due to the interlocking of the mechanics, the recuperation potentials, which usually can be found in lifting applications, could not be used. The efficiency of the system in this application is significantly influenced by the way it is mechanically realized.

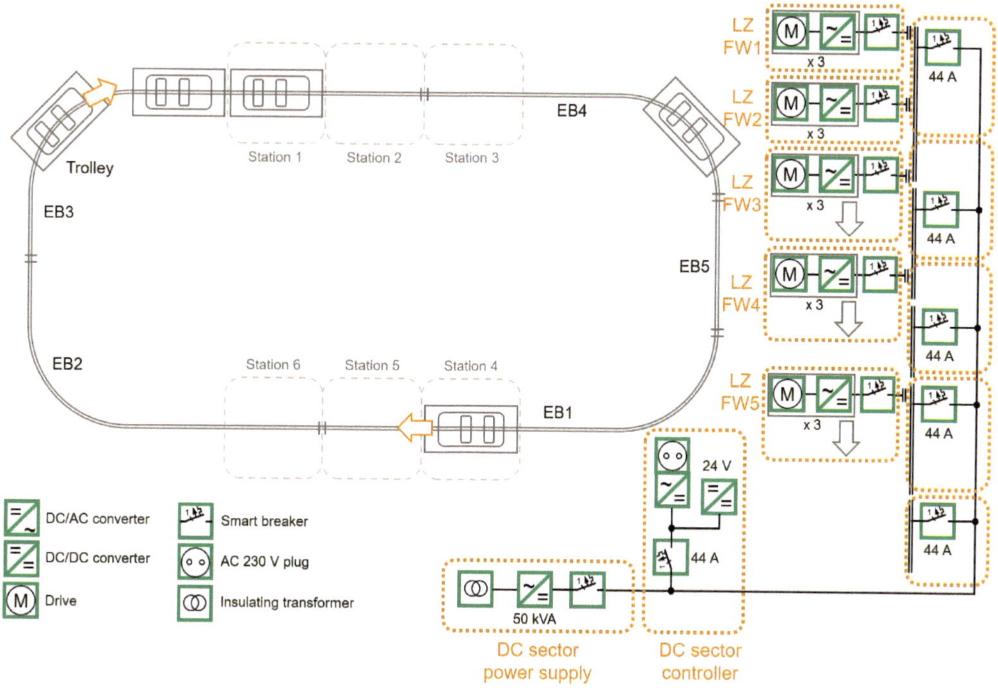


**Figure 9.8** Structure of the DC grid with all connected participants of the electrified monorail system

## ■ 9.3 Body Shop

The third application area is body shop work with dynamic industrial robots and process technology. This context was addressed with the fourth test environment. The respective application is a replica of a real production cell in the automotive industry, consisting of four industrial robots. In addition to gluing and riveting, it was spot welding that was of particular interest from an energy perspective. Spot welding requires a current of several 100 A for very short periods of a few milliseconds. To smooth these spikes, additional capacitors and a flywheel mass storage device were integrated into the DC grid. The application had to demonstrate how the DC grid handles such power pulses. In addition, the robots, which until now have only used the recuperated energy internally in the servo axis network, were to exchange the energy. Two lithium storage units were integrated to bridge longer power outage periods. Last, but not least, the question was addressed, to what extent there was a possibility of integrating renewable energy in the form of a PV system into the DC grid. Figure 9.9 gives an insight into the production cell.

**Figure 9.9** DC-powered robotic cell in the body shop

The entire body-in-white cell was supplied by a central Active Infeed Converter. The individual processing stations were each combined in a DC sector. The welding itself was separated into its own DC sector in order to be able to design the protection technology according to requirements. The individual suppliers and storage facilities were also structured into individual DC sectors and correspondingly protected separately with Smart Breakers. The characteristic control of the grid is performed on the AIC, the DC/DC controller of the PV system and the DC/DC controllers of the batteries and the flywheel mass storage system. The capacitor was directly connected to the DC grid without regulation. The robotics and process equipment were also not integrated into the grid management. These loads were controlled by the production process and should not be influenced by the energy situation in the DC grid. Figure 9.10 shows the structure of the DC system of the body shop cell. It is clear that the utilities and storage units are connected to the power distribution system of the busbar and are thus quasi external to the production cell as an energy system. The robot cell itself is connected to the energy supply system via a distribution cabinet, as is common in automotive engineering. The outgoing feeders for the robots and the process technology are arranged in a distribution cabinet. The DC-powered cell thus follows the structural principles of the automotive industry for energy distribution.

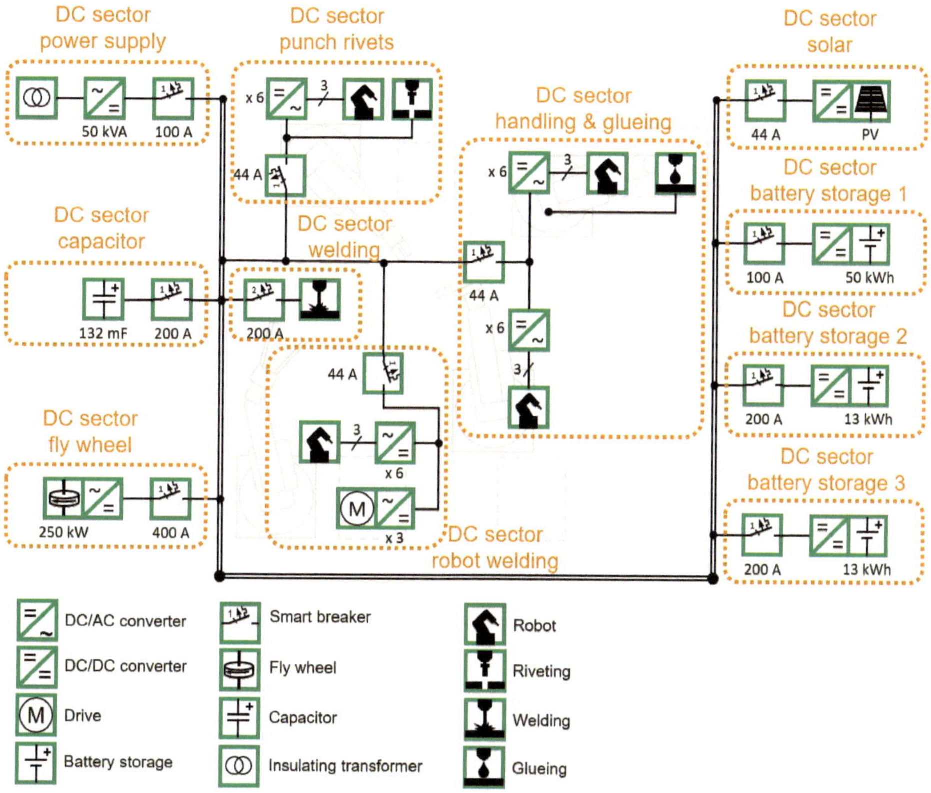


**Figure 9.10** Structure of the DC grid with all connected participants of the robot cell

The body-in-white cell can be operated without problems even with the dynamic welding pulses from a common DC grid. This is the most important finding. Measurements show that sufficient power is provided by the storage units at the moment of welding, so that the voltage in the DC grid remains within the normal operating range. The neighboring processes could thus always continue to operate undisturbed. The central supply via the AIC was dimensioned so small at 50 kW that it alone could not have provided the necessary power of over 200 kW at peak for welding. Nor did it have to. Right at the first moment, the capacitor provided additional power, compensating for the minimum lag time of the flywheel mass accumulator. This power storage device then subsequently fed the required power into the DC grid. The DC grid remained stable in its power and energy balance. This also demonstrated that the concept of characteristic control in different active participants provides a stable grid without additional communication infrastructure between the participants. The storage units were functionally prioritized by their characteristic curves, allowing the lithium storage units to absorb excess energy and replace the AIC in case of failure. The flywheel mass storage remains unaffected by such rather minor balancing operations and only intervenes in situations of extremely high power demand. If the power drops into the normal range, the flywheel mass accumulator immediately recharges itself to be ready for the next welding impulse.

It was possible to handle these dynamics and the varying power ratios between the DC sectors well with the Smart Breaker as the fuse component. Short-circuit tests showed that the requirement for selectivity was also met in most cases (cf. Chapter 7.2).

In addition, it was shown that mixed operation between hybrid and electronic Smart Breakers on one level is also possible. However, not only electrical safety could be ensured in the plant, but also electromagnetic compatibility (EMC). Measurements in the hall and regarding the DC grid showed that the limit values derived from the various standards were complied with.

With regard to concrete savings, direct comparisons with an AC system are lacking. However, it can be stated that the connected load for the entire body-in-white cell was significantly reduced by the fast power storage units. Compared to the first DC realization of this robot cell in the EU project AREUS [Pell2015], the supply power of the AIC could be reduced from 450 kW to 50 kW and thus by a factor of 9. The braking energy of the robots, which according to studies in other work [Meik2011] can be up to 10%, has remained completely in the system. The coupling of the suppliers and storage units could be realized with low effort without a higher-level management system, so that these costs are eliminated in comparison to AC systems.

## Literature

[Meik2011] Meike, Davis; Ribickis, Leonids: Recuperated Energy Savings Potential and Approaches Robotics. Proceedings of IEEE International Conference on Automation Science and Engineering, IEEE, Trieste, 2011

[Pell2015] Pellicciari, Marcello; Avotins, Ansis; Bengtsson, Kristofer; Berselli, Giovanni; Bey, Niki; Lennartson, Bengt; Meike, Davis: AREUS – Innovative Hardware and Software for Sustainable Industrial Robotics. IEEE, Gothenburg, 2015

# 10 Summary and Outlook

Alexander Sauer
Holger Borcherding
Hartwig Stammberger
Karl-Peter Simon
Timm Kuhlmann

*The examples have shown: Industrial DC grids work and save resources.*

The preceding chapters have detailed the technical concepts, that were necessary to realize the four, prototypical DC grids with real machines and plants (cf. Chapter 9). In addition to the electric drives and process peripherals such as welding, all four application cases were equipped with energy storage systems integrated into the plants. With this, the foundation for an implementation of a DC factory has been established. The potential for reducing energy demand, for making energy supply more flexible and for reducing peak loads, thus making energy supply more stable, has been demonstrated. In numerous tests, the proper functioning of each system was shown. As described in chapter 9 on the basis of the test plants, the braking energy from dynamic processes can be fed back and used directly in the DC grid. Only the difference in energy between the required motor energy and the regenerated braking energy has to be fed in. In combination with storage units for smoothing short-term load peaks, it has been shown that the connected load of a plant can be reduced by 85%. In addition to equipment costs for converters, line filters and supply devices, energy costs could also be saved via the charge for the power provided. Energy consumption can be reduced by 6% to 8%. As explained in Chapter 2, this energy saving leads to a $CO_2$ reduction of 360 kg $CO_2e$[1] per year for a machine tool, for example. In addition, further resources can be saved. A case in point is the elimination of cores in a 3-wire system which reduces the copper requirement of a DC cable by approx. 40% compared to a 5-wire system. In addition, up to 25% of the insulation material is saved. This leads to a total reduction of up to 38 kg $CO_2e$ per km of cable with a cross-section of 2.5 mm$^2$. Electronic components can be saved by eliminating numerous rectifiers in the drive controllers and reducing the central feed-in power. For a machine tool, an additional 230 kg $CO_2e$ can be saved.

Photovoltaic systems, combined heat and power plants or other forms of decentralized, electrical energy generation are increasingly being integrated into local factory grids to generate electricity and reduce the $CO_2$ impact of their own produc-

[1] $CO_2e = CO_2$ equivalent

tion. The energy produced in this way can be fed into a DC grid more efficiently than if converted for an AC grid, since direct energy balancing is possible without AC/DC conversion. Energy-saving LED lights and all electronic control devices and computers require DC voltage. The number of DC voltage consumers will increase even further with increasing digitalization. Many decentralized conversions from AC to DC will be eliminated, and the resulting grid feedbacks will be significantly reduced. This change supports both an economic energy turnaround and the global reduction of $CO_2$ emissions by significantly increasing energy efficiency.

The investment costs for a DC power supply with a developing component market are roughly equivalent to today's AC power supply systems in factories. Due to rapid developments in the electronics and semiconductor industry, a cost advantage may develop in the future, especially for more complex systems with different power generators and storage systems.

*A DC grid is the consistent implementation of "Industry 4.0" and "Smart Grid" in the factory.*

If, in the future, electric motors are operated with DC-powered frequency converters as standard, then it will be possible to always integrate the drive into a modern control environment, as required by "Industry 4.0" applications. This means that both status and control signals can be transmitted, such as electronic nameplate data, current energy consumption data, load cycle and much more. By using DC grids, all electric drives become "Industry 4.0" capable. This capability leads to functions such as autonomous tuning for easy commissioning, oscilloscope function for recording signals in real time, or energy management for retrieving relevant consumption data. In this way, information about the condition of the machine and about an optimization of its operation can be implemented.

During its life cycle, a drive goes through various stations in the value creation process. From planning and production to integration in machines, commissioning, use in the production process and finally recycling. In the course of the digital transformation of development as well as production and life cycle processes, a uniform, manufacturer-independent provision of data plays a central role. A uniform data structure in accordance with "Industry 4.0" semantics is now available for this purpose. It was defined in the *eCl@ss* classification system, which in turn is based on international standards. This provides standardized data as the basis for creating information models complying with the OPC UA protocol.

The digital transformation of the energy supply grid leads to a *smart grid*, a microgrid that can be operated independently and that exchanges energy with the higher-level supply grid or other microgrids. The energy supply of the DC factory is based on such a microgrid. Different local energy generators and energy storage systems supply the factory. They ensure an efficient, robust and stable energy supply. At the same time, the smart grid interacts with the supply grid. Disturbances are decoupled from one another. It is also possible to react flexibly to the regional or national energy market. In the event of a surplus in the utility grid, energy can be absorbed in the DC plant's storage facilities. When energy demand in the utility grid is high, the smart grid of the DC factory feeds back energy that is not needed. This grid serviceability supports the public grids and contributes to the success of the energy transition. In addition, energy purchase can be decoupled from the actual energy demand, so that economically optimized energy can be traded. The DC grid offers an ideal basis for the energy control of the smart grid participants through an inherent characteristic-based control.

*The DC grid unlocks the potentials of an intelligent energy grid by introducing plug&play capability.*

While plant technology benefits from simplification, modern energy supply in companies requires increasing control of the complexity of generators, consumers and storage systems. On the part of the energy supply market, flexibility is demanded from companies in their role as energy consumers - or at least will be increasingly rewarded in the future. In this way, the grid operator can balance the increasing amount of fluctuating regenerative energy sources in the grid and make them controllable. These complex energy supply tasks are increasingly shifting to the factory energy grid. The factory grid operator, i.e., the respective company itself, has to also manage the balancing of the energy sources as well as the securing of the supply in an economically optimal way. In a changing production environment, which is subject to ever greater energy fluctuations driven by individualization and rapid product changes, strategies are needed to adapt the energy system rapidly. Components such as storage and generators must be integratable quickly and easily. The complex and manual integration of the control of such new participants must be automatic and self-learning. The decentralized grid management of the DC grid already provides a good basis for this. In the future, the subordinate storage control should tie in with this. It should independently control the optimal operating point of the connected storage under the influence of external fluctuations. Machine learning methods are suitable for this purpose. In this way, developments in the energy curve of the system can be anticipated. As a result, storage systems can be designed smaller and operated more efficiently. In addition, the aging of components can be detected, which leads to other participants gradually taking over the task or to such components being replaced in time. Disturbances in the factory's energy system are avoided. With respect to this topic, too, the DC factory makes an important contribution in that it ensures a more stable and resource-saving energy supply in factories.

*The transformation process from today's energy supply to a DC factory can either be top-down or bottom-up.*

The system concept with the specifications for devices and grid management as well as the deliberations on safety show that the DC factory is not a vision but is already becoming reality. The first approaches of an adapted planning and design concept for the energy supply in the DC factory address the new tasks that companies will face in the context of infrastructure planning and system planning. Upstream of the task-oriented view of the direct implementation of a DC grid in the factory, a strategic approach to the transformation of the electrical power supply in the factory is required. This transformation process starts with today's power supply based on AC voltage and ends with the DC factory. Obviously, the specifics at the start of such process and especially the environment of the respective company have to be considered. In addition, the advancing technical maturity of the system components must be taken into account in the factory transformation process. Since the factory infrastructure has a much longer life cycle than production technology has, they should be considered differently. In very simplified terms, the factory infrastructure has a supply character. Among other things, it is responsible for the stable, robust, efficient, ecological and economical provision of electrical energy for the production processes as well as the electrically supplied infrastructure such as ventilation, lighting or even compressed air. The hall infrastructure also includes the integration of decentralized self-energy supply systems into the energy system of the factory. When planning the hall infrastructure for new factory buildings – greenfield planning – a (partial) DC supply can already be considered today. For example, Daimler is installing a DC grid in the new "Factory 56" building – the "car factory of the future". It efficiently connects part of the PV system on the roof of the building with a battery storage system and thus supplies parts of the building infrastructure as well as charging stations for electric vehicles. This implementation is already possible today, since due to the large outputs and energy quantities with rather few but energetically versatile grid participants, an overlapping DC grid already brings important advantages. At a later stage, production can still be supplied in an intermediate step with grid-forming DC/AC converters, as known from the photovoltaic industry, or conventionally by means of parallel AC infrastructure. If a coupling between the AC grid and the DC grid takes place, the production is robustly supplied from different self-owned sources such as PV systems or storage systems. Disturbances from the utility grid are filtered out by the DC grid. The same applies to the grid feedback from the controlled drives on the supply grid. Due to the small number of participants in a transformation of the hall infrastructure that is initially very focused, technical solutions can be used that are not yet fully optimized for larger series applications. Additional costs, for example in the area of protection technology, have only a small impact on the total investment volume due to the small number of DC grid participants to be connected. On the other hand, the factory's power supply is equipped for the future. It can be flexibly expanded and adapted to further developments in the area of storage and generation technologies, to changes of grid operator requirements or to changing conditions in the energy environment of production.

If an energy supply already exists in the production hall, it must be carefully examined to what extent a parallel new DC infrastructure would justify an additional expenditure. If strategic investments in storage systems and energy self-generation plants are planned, DC supply may already make sense in the first development stages of this strategy. If the goal is only to integrate an isolated component into the power system, such as a PV system, the DC grid is unlikely to deliver the key benefits of energy exchange and conversion loss reduction to the point where the higher investment would be justified. Brownfield planning of a DC-based indoor infrastructure as a selective upgrade or expansion of the existing AC infrastructure will become widely adopted in a variety of use cases as the stable supplier market for DC components increases.

Parallel to the transformation process of the hall infrastructure, which represents a kind of top-down development approach, the question arises as to how the energy supply in machines and plants will develop. The DC grid within the plant interconnects numerous drives that are grouped and protected in individual DC sectors. While the converter technology is ready for use, as mentioned, efforts are still needed for marketable safety technology. When this is ready, plants can internally supply the power for loads with a DC grid. The connection to the hall infrastructure is made via decentralized AC/DC converters from the AC grid that is still available in most cases. In many companies, the modernization of plant technology is taking place step by step. Old machines are gradually replaced by modern machines. Thus, one DC machine after the other is moving into the companies. Several DC machines can be operated in an energetic network in which the internal DC grids are coupled with each other. A DC grid is slowly and gradually being created. Especially the initially decentralized supply technology would benefit, since the simultaneity factor means that the sum of the powers of the supplying AC/DC converters can be designed lower than the sum of the connected powers of the DC machines. With the DC grid growing on the shop floor, elements of the hall infrastructure are integrated in a next step. A PV system or an electrical storage system could be coupled with the DC grid. From this, the DC grid in the factory continues to grow from the bottom up to the point where the benefits of the size of the interconnection of the DC grid justify the conversion costs of the still existing AC infrastructure. This outlined development can be called a bottom-up approach, as the DC grid expands in the factory from the bottom.

Which of the transformation processes will take place depends on the modernization strategy of the respective company and on the supply situation for DC technology. This must be available not only on a component basis, but also in form of equipped machines and systems. Particularly for machine and plant construction, there are further design options. Reducing the size of a converter by eliminating the rectifier leads to more compact devices due to smaller heat sinks and fewer electronics, which in turn makes it easier to integrate such components directly into a motor's terminal box. Decentralization of power electronics in mechanical engineering can lead to smaller control cabinets and simplified modularization of systems. There is also the simplification of cable routing, as fewer motor cables need to be routed. Machine design can also benefit at this point. Despite being connected to the AC grid with a rectifier, even a machine supplied internally with DC voltage can have advantages that go beyond pure electrical engineering and energy savings. Changes can be expected in this market environment in the future.

*An industrial DC grid has further energy-saving potentials in the context of modern drive technology.*

Since previous legal requirements for increasing energy efficiency in drive technology have not had the success that was hoped for, a rethink is urgently needed. This is essentially due to the fact that the previous regulations only specified the electric motor for grid operation. However, in order to economically design an application for energy efficiency, it makes more sense to optimize the entire drive train. If an electric motor is optimized for grid operation at a high additional material cost and only for the rated operating point, then it will rarely be used. Many drives are used in partial load operation and in more than one third of the applications the speed is also changed electronically via an inverter. However, the social and also political demand is to optimize the real operation of an application in order to really save significant energy. Furthermore, this objective should be implemented as ecologically as possible. With the minimum efficiency required by law today (usually IE3), considerably more material has to be used, which of course has a negative impact on the $CO_2$ balance of the motor. Thus, it may make more ecological sense to optimize drives that have only a very short operating life in a different way, since a lower energy requirement also results in lower potential savings. This is because more $CO_2$ emissions are then caused for the production of an IE3 motor than can be saved due to the lower power consumption in the comparatively short operating time of the motor.

With the introduction of a DC grid in the factory, the opportunity arises to also optimize electric motors for operation with frequency converters. For more than 130 years, this has been optimized for AC grid operation and also regulated by law. The use of converters, which is common today, for electronically changing the motor speed, was largely ignored. On a DC grid, however, all asynchronous motors are systematically operated with frequency converters, since an asynchronous motor requires an AC voltage with variable voltage and frequency. It cannot be supplied directly with DC voltage. In order to take full advantage of operation with a frequency converter, the motor winding should no longer be optimized for 50 Hz mains supply, but for operation with the converter. This allows the motor to be operated above 50 Hz. With the higher frequency, the efficiency and the rated power of the asynchronous motor increase. In addition, the higher frequency enables a larger speed setting range.

Optimizing the asynchronous motors for operation on the converter can therefore both increase energy efficiency and save material (e.g., copper). The higher the rated frequency of an electric motor, the more material can be saved. A standard frequency converter can very easily provide a higher frequency. To accelerate this development, the legal framework must be expanded in the direction of the overall balance of the entire drive system with all connected components, such as the line filter, frequency converter, motor, braking resistor, cable connections, gears, etc. Thus, the central DC power supply offers the opportunity to open up new possibilities for higher resource and energy efficiency. By operating all electric motors with a frequency converter, in principle every drive becomes variable in terms of speed. New opportunities arise for increasing the efficiency of the electric drive system in conjunction with the mechanical system and process operations. There are over 8 billion electric motors in use in the EU, consuming half of the electricity produced [EC2019]. Annually, about 750 million motors are sold, with energy consumption of about 2000 TWh per year and emissions of about 800 Mt $CO_2$ (equivalent per year). Thus, energy efficiency improvements of electric drives lead to significant overall energy savings with further benefits also in terms of EU greenhouse gas emission targets and supply reliability. In industrial applications where the motor is used intensively, energy costs can be more than hundred times the acquisition costs.

*Frequency converters are ready for use and have to become interoperable.*

The technology for building the DC grid is well known and many components are already commercially available. For example, frequency converters with DC connections for coupling DC links are state of the art. Many manufacturers already offer solutions in control cabinet devices. For DC grids, often only minor or no adaptations are necessary. The fundamental energy-saving potential of multi-axis converters with DC coupling compared to single-axis converters is also frequently cited in company publications as a particular advantage. It is therefore not surprising that the initiative for the DC-INDUSTRIE research project came from the "Electrical Drives" working group of the ZVEI[2], in which the manufacturers of frequency converters are organized. There is still little supply in the case of switchgear cabinet-less, decentralized devices, but due to the major constructional advantages resulting from the elimination of the rectifier and braking resistors, fewer terminals and capacitors, and thus much easier integration, there is a significant market de-

[2] ZVEI is the German Electrical and Electronic Manufacturers' Association (Zentralverband Elektrotechnik- und Elektronikindustrie).

mand here in particular. Ultimately, the DC factory concept is expected to prepare the market for highly efficient synchronous and asynchronous motor technologies, enabling energy savings potentials in excess of 30% that would be impossible with today's grid-fed asynchronous motors.

*DC protection and switching technology will continue to develop rapidly.*

The use of widely available components for switching and protection technology has so far been limited. For example, DC switches are known and also in use, e.g., in railroad technology. However, for applications in DC grids with many energy storage devices such as capacitors and batteries, it is necessary for the breaking times of the circuit breakers to be significantly shorter than those of the electromechanical circuit breakers normally used in AC grids today. A key task for switchgear manufacturers is therefore to further develop power electronic switchgear that can switch much faster due to the principle involved, since they do not have to move any – or in the case of hybrid switches only small – masses. The wide range of applications for power electronics also makes this technology attractive for switching. There are many activities to reduce the power loss, which is due to principle higher than that of mechanical metal contacts.

The "Smart Breaker" component described in this book comprises the protective functions required for a safe operation of the DC grid. These technical approaches are being further developed, resulting in compact and marketable products.

Selectivity between individual protective devices, which ensures that only the part of the system affected by a fault is switched off, is an important user requirement for both DC and AC. In many cases, it can already be ensured, as described in chapter 9. However, just as with AC, it is not guaranteed in every case with every combination of breakers. The fast breaking of power electronic switches eliminates the possibility of signal-based selectivity between breaking devices. The focus of further developments must therefore be on evaluating the current and voltage signals in the breaker itself.

*DC/DC converters and supply devices (AIC) are core components.*

The availability of series components is also still limited as far as connecting storage systems and energy generators are concerned. For the economic integration of self-generated energy and storage systems, DC/DC converters are required in series quantities in a wide performance range. Currently, there are only a few suppliers on the market, most of them with only one product. The functional, size and cost advantages of fast-switching wide bandgap power semiconductors have not yet been realized.

The range of supply devices is considerably wider. The system concept of the DC grid (cf. Chapter 5) details two principal possibilities of controlled and uncontrolled DC rectification with DC-side or AC-side grounding. Unregulated rectifiers have been used for decades in all power ranges. For regulated power supplies, AIC technology has been established for many years in electric drive power supply modules and solar converters. However, the use of unshielded DC lines, which – from an EMC point of view – may only have low interference voltages, is only possible with potential separation or with a considerable filtering effort for the basic AIC circuitry used to date.

Topologies of AIC with integrated potential separation, high switching frequencies for lower material requirements and simple parallel connectivity have already been addressed in many academic papers and some pilot installations, but markets with high volumes have been lacking so far. DC grids now provide the basis for such innovative technologies increasingly put forth powerful and cost-effective supply devices. For example, the isolating transformer still required in the system concept is possibly only a bridge that will be eliminated with further optimization of the supply system technology. In addition to energy efficiency gains from transformer loss, cost reductions are also an important step toward the DC factory.

Other requirements that will be very important for future supply devices are reliability, a long service life and short-circuit resistance, especially with central grounding. These basic requirements for components in main branches of the power flow are already known from high-voltage DC transmission. Short-circuit strength can be achieved by oversizing, but also by current-limiting AIC topologies. For high reliability, extensive qualification measures and experience are necessary. In all these aspects, it is expected that artificial intelligence can be applied to enable condition monitoring and predictive maintenance to increase availability. It is obvious that the DC factory goes hand in hand with "Industry 4.0" approaches.

*DC grids have no negative grid feedback on the utility grid.*

The use of AIC as supply devices eliminates current harmonics in the AC supply grid, with which standard AC-AC frequency converters massively burden the grids. This has many advantages. There is no need for compensation systems, since the power factor can be controlled to "one". AC equipment such as transformers can be utilized to a greater extent, since significantly more active power can be transmitted due to the lack of reactive power. No harmonic filters are required in the AC grid to protect other equipment from distorted line voltage.

If required, an AIC can provide fundamental reactive power or filter current harmonics on the connection side of the AC supply grid. By capping load peaks, voltage drops in the AC grid decrease, increasing in turn the stability of the AC grid. However, all this requires advanced AICs consistently designed to meet the requirements of the DC factory.

The overall advantages of DC grids become particularly apparent when considering the system. Cost and benefit advantages through load peak reduction, robust and stable energy supply for production and use of self-generated energy require planning efforts. Therefore, the system-technical properties of the DC grid with their advantages must be directly accessible to the planners. Planning begins with the selection of the right supply and grounding concept.

*DC grids first establish themselves in factory automation.*

A large number of electric drives and thus dynamic movements are particularly found in the process industry and factory automation. The chemical industry within the process industry has many additional power supply requirements that are tied to the specific characteristics of the AC supply. Complex, sensitive or firmly established processes can only be modernized at great expense. In this sector, deployment of extended DC grids will still require additional research and effort. The situation is different in other areas of the process industry, such as wastewater technology, sewage treatment plants and waterworks. Long operating times of the drives with high demands on process stability, few additional requirements on the control and supply technology, and proximity to power generation plants are ideal conditions for benefiting from the potentials of an open DC grid.

The potential applications in factory automation are similarly well assessed. DC grids can initially be used as a supplement in existing production plants. A prerequisite for worldwide dissemination is the broad availability of DC components and an international standardization. There are already a large number of small, decentralized DC grids in motion control applications. However, these are mostly manufacturer-specific solutions, which therefore cannot be operated with third-party components as an open standardized DC grid. Only when it is possible to connect company-independent components in an open DC grid, the conditions for acceptance are met. Standard components from different manufacturers must be available worldwide. Acceptance among end users should be equally international. Mechanical engineering companies that supply entire production could operate them on a DC grid already today. This would enable considerable cost reductions, especially if a large number of drives have to be equipped with electronic speed control. The conversion from AC to DC voltage takes place centrally only once and the decentralized converters can be simplified, since only one converter is required. Additional braking resistors in conjunction with frequency converters for dissipating braking energy (hoist drives, etc.) are no longer required. This provides the opportunity to reduce system costs and increase energy efficiency. Companies that look into the possibilities of introducing decentralized DC grids at a very early stage can thus prepare for these changes in time and thereby secure their market position. The concept of the DC factory provides many convincing arguments for the introduction of DC grids, so it is essential to push the introduction. The implementation of DC grids has long since begun in machine solutions (motion control). This is continued in production plants with open standardized DC grids and will finally lead to a cross-factory DC grid. Since the most experience and also the largest component availability is currently given with respect to material handling applications, open DC grids could very soon be realized for such applications. Industrial sectors that already equip almost all asynchronous motors with frequency converters will be the first to introduce DC grids. Predestined for this are production plants for automobile production, intralogistics and plants for food and beverage production.

*The foundation for the worldwide standardization of industrial DC grids has been established.*

The advantages of the DC factory and its further development have been discussed in detail in the previous sections. This means that a DC grid based on the system specification presented is basically ready for use. Furthermore, the cross-manufacturer function has already been demonstrated. A broad acceptance has been achieved due to the high number of participating companies in the realized applications. Components can be seen being explicitly placed on the market with reference to DC-INDUSTRIE.

As a consistent next step, these results should drive standardization at the national and international level forward. Standardization creates further acceptance for DC grids in factories. This concerns the DC voltage level, the definition of grounding and protection concepts, and much more. Only in this way can all components within a DC grid work together without interference. At the same time, it protects investments by providing the security of being able to expand and flexibly redesign the power supply system in the future. For component suppliers, development investments are safeguarded by standards. This creates confidence that the developments are sustainable and will be refinanced by the proceeds from the sale of components. Only this security – for users and suppliers alike – will lead to the widespread implementation of DC factories.

Since the system specification describes an industrial DC grid holistically both at the component level and at the system level, a consistency and worldwide applicability can be achieved which – for historical reasons – could never be achieved with AC grids. This unique opportunity is to be used. Since the realization partners involved so far are active worldwide, the DC factory concept is well on its way to becoming the basis for internationally standardized industrial DC grids.

*Conclusion*

We have looked back to the point where DC technology was competitively inferior to AC technology. The world has moved on and it was time to ask whether the developments of the last 130 years and especially that of the last decades in the field of power electronics, energy supply and our current demand for a climate-neutral economy as well as the associated need to increase energy and resource efficiency necessitate a change also in the industrial electrical energy system. The DC factory is the answer – energy-efficient, robust, forward-looking. It forms a stable energy supply for production processes that have become more sensitive and which the public grid is often no longer able to provide due to the increasing feed-in of renewable energies. It contributes to climate-neutral production by increasing power efficiency through the elimination of conversion stages and recuperation, by efficiently integrating locally renewable power generators, and by reducing the resources required for the factory's energy infrastructure.

Factory operators who are already today transitioning to DC factories will quickly be able to realize comparative cost advantages through DC technologies. For factory equipment suppliers, it is essential to take a close look at the new possibilities of DC technology in order to be competitive in the future. Today, we are still at the beginning of a steep learning curve that will determine competitive advantages and future standards. The DC factory is the invisible and silent revolution in the factory. After 130 years of AC dominance and climate-damaging industrialization, a fresh wind is blowing toward the green sustainable factory – with direct current.

## Literature

[EC2019] European Commission: "COMMISSION STAFF WORKING DOCUMENT IMPACT ASSESSMENTCOMMISSION REGULATION (EU) …/… laying down ecodesign requirements for electric motors and variable speed drives pursuant to Directive 2009/125/EC of the European Parliament and of the Council", 2019, *https://ec.europa.eu/transparency/regdoc/rep/10102/2019/EN/SWD-2019-343-F1-EN-MAIN-PART-1.PDF*

# Index

## S



## T



## U



## V



## W